D1562739

Green Energy and Technology

For further volumes:
http://www.springer.com/series/8059

Stefan Emeis

Wind Energy Meteorology

Atmospheric Physics for Wind Power Generation

Stefan Emeis
Institut für Meteorologie und Klimaforschung
Karlsruher Institut für Technologie
Kreuzeckbahnstr. 19
82467 Garmisch-Partenkirchen
Germany

ISSN 1865-3529 ISSN 1865-3537 (electronic)
ISBN 978-3-642-30522-1 ISBN 978-3-642-30523-8 (eBook)
DOI 10.1007/978-3-642-30523-8
Springer Heidelberg New York Dordrecht London

Library of Congress Control Number: 2012940943

Printed on acid-free paper

Springer is part of Springer Science+Business Media (www.springer.com)

Preface

Many books have already been written on converting the kinetic energy of the wind into mainly electrical energy. This one, written by a meteorologist, will entirely concentrate on the atmospheric features and phenomena influencing the generation of electric power from the wind. Such a book is presently—to my knowledge—unavailable. This book presents part of what is today called 'energy meteorology', a presently emerging new sub-discipline in the field of meteorology.

I thank Springer Science Media for the invitation to write such a book which is designed to fit into the series "Green Energy and Technology" which deals with various aspects on renewable energies. This series already comprises several titles on wind energy. Once again most of these titles are on technical aspects but none of these concentrates on the meteorological boundary conditions for the conversion of energy from the wind. My special thanks go to Claus Ascheron of Springer who accompanied the preparation of the manuscript and gave invaluable advice.

I am working as a scientist in the discipline of meteorology since the 1980s. The field of energy meteorology has found my attention for more than 20 years, although the term 'energy meteorology' is much newer. My interest in this subject was initiated during a sabbatical leave at the Wind Energy Institute of the Danish National Laboratory (today part of the Danish Technical University, DTU) at Risø near Roskilde, Denmark. Here, I met boundary-layer meteorology experts and saw one of the first test sites for wind turbines. Essentially, wind energy meteorology is a special section of boundary-layer meteorology. I still have very fruitful and friendly contacts with this renowned Danish research institute. In 1991 in Risø I also met the late Sten Frandsen for the first time. Discussions with him started my attention to the wind park issue. What is presented here in Chap. 6 in this book is a much more elaborated version of an idea which was born during that first stay in Risø. Thus, I dedicate Chap. 6 to him.

Later I worked many years on acoustic profiling of the atmospheric boundary layer with SODAR devices. These instruments allow for a surface-based detection of the boundary-layer wind profile based on an analysis of the Doppler shift of the backscattered signal. This is a technique, which captured the interest of the wind energy community in the 1990 s. In recent years, my experimental activities and expertise have been complemented by the operation of ceilometers, RASS and wind lidars. In addition, I focussed on the investigation of peculiarities of the marine boundary layer from data from the German offshore measurement platform FINO1 in several research projects. I am a member of the Southern German wind energy research alliance WindForS.

The marine boundary layer projects have been funded through several grants by the German Ministry of the Environment, Nature Protection and Nuclear Safety (BMU, FKZ 032 99 61, 032 50 50, 032 53 04). These projects within the RAVE program (Research at Alpha Ventus) were initiated in order to accompany scientifically the establishment of the first German offshore wind park Alpha Ventus, which is situated in the German Bight roughly 45 km away from the nearest coast in about 30 m deep waters. Six years before the first turbine installation began, a 100 m meteorological measurement tower (FINO1) was erected at the later site of Alpha Ventus in order to facilitate the studies of the marine boundary layer. Much of the information concerning the marine boundary layer presented in this volume is based on data obtained at this tower which has eight measurement platforms between 30 and 100 m. The evaluation of this tower data has mainly been performed by two PhD students of mine; Matthias Türk[1] and Richard Foreman[2]. Further funding is available through a project lead by Sven-Erik Gryning from Risø DTU and which is presently supported by Forsknings—og Innovationsstyrelsen at the Danish Ministeriet for Videnskab, Teknologi og Udvikling (Sagsnr 2104-08-0025) within the project: "Large wind turbines—the wind profile up to 400 m". The results for urban boundary layers are partly based on studies funded by the German Ministry of Education and Research (BMBF) in the framework of the AFO2000 program. The data from Graswang in Figs. A.1 and A.2 in Appendix A have been obtained in the framework of the TERENO programme of the Helmholtz society funded by the BMBF. The studies on flow over complex terrain have partly been made possible through the financial support by several private enterprises.

A draft version of the manuscript has been read by Beatriz Cañadillas, Richard Foreman, Tom Neumann, and Matthias Türk. I thank all of them for their valuable suggestions, help and advice. Nevertheless, it is me to be blamed for any errors or inconsistencies. I hope that this book will help to bring the meteorological part in wind power conversion to a better visibility. We urgently need efficient strategies to generate renewable energies for the energy demand of mankind and a better understanding of the meteorological prerequisites for wind power generation should be part of this strategy.

Spring 2012 Stefan Emeis

---

[1] Türk, M.: Ermittlung designrelevanter Belastungsparameter für Offshore-Windkraftanlagen. PhD thesis, University of Cologne (2009) (Available from: http://kups.ub.uni-koeln.de/2799/)

[2] Foreman, R.: Improved calculation of offshore meteorological parameters for applications in wind energy. PhD thesis, Faculty of Mathematics and Natural Sciences, University of Cologne (2012)

Contents

List of Variables

a	reduction factor
	constant in stability correction function
	power-law exponent
b	constant in stability correction function
c	wave age
c_p	specific heat
c_{ph}	phase speed
c_2	factor in Eq. (3.91)
c_s	surface drag coefficient
c_t	turbine drag coefficient
c_{teff}	effective drag coefficient
d	displacement height
f	horizontal Coriolis parameter
f^*	vertical Coriolis parameter
$f(x)$	Weibull function
g	gravity
h	height of atmospheric layers
k	wake decay coefficient
k_{max}	maximum wave number of spectrum
l	mixing length
	inner layer height (flow over hills)
m	mass-specific momentum
n	frequency
p	air pressure
q	specific humidity
r	radius of curvature
s	distance
t	time
u	west–east wind component

u_{10} 10 m wind speed
u_g west–east component of geostrophic wind speed
u_h hub height wind speed
u_{pot} wind speed from potential flow theory
u_{gust} gust wind speed
u_* friction velocity
u_∞ upstream wind speed (flow over hills)
v south–north wind component
v_1 maximum wind speed with 1 year return period
v_{50} maximum wind speed with 50 year return period
v_{e1} 10 min extreme wind speed with 1 year return period
v_{e50} 10 min extreme wind speed with 50 year return period
v_g south–north component of geostrophic wind
w vertical wind component
w_* convective velocity scale
x eastward pointing coordinate
abbreviation in stability correction function
y northward pointing coordinate
z vertical coordinate
z_0 roughness length
z_i boundary-layer height
z_r reference height
z_g Ekman layer height
z_p Prandtl layer height
z_m height variable in shape function
z_A height variable in shape function
A scale factor in Weibull distribution
A_g geostrophic scale factor in Weibull distribution
A_r rotor area of a wind turbine
B Bowen ratio
B_i backscatter profile
BR buoyancy ratio
C_D drag coefficient
C_{DN10} neutral drag coefficient at 10 m height
C_T thrust coefficient
D distance
D_u ageostrophic wind component
D_v ageostrophic wind component
D_w wake width
E_{wind} wind energy
F Weibull distribution function
$F_{x,y,z}$ components of turbulent friction
G gust factor

$G(p)$	percent point function
H	height of hills
H_s	significant wave height
I_u	turbulence intensity
K_M	turbulent exchange coefficient/turbulent viscosity
L	half width of hills
L_v	latent heat of condensation
L_x	integral length scale
$L_{L,M,U}$	length scales in wind profile law
$L_{u,v,w}$	length scales in spectra
L_*	Obukhov length
M_n	higher moments of Weibull distribution function
P_0	pressure function (flow over hills)
P_δ	pressure function (flow over hills)
R	gas constant
R_u	reduction factor
R_t	reduction factor
R_n	reduction factor
$S_{u,v,w}$	spectra
Sf	structure function
T	wave period
	temperature
	time scale
	stability correction function in wind profile law
T_m	mean temperature
T_v	virtual temperature
α	exponential factor
	Charnock parameter
α_0	angle between geostrophic and surface wind
η	limiting value for mixing length
φ	latitude
	wind turning angle
φ_ε	dissipation rate for kinetic energy
ϕ	differential stability correction function
γ	inverse length scale in Ekman layer wind profile
κ	von Kármán constant
ρ	air density
σ	shape function for hills
$\sigma_{u,v,w}$	standard deviations of wind components
σ_{u90}	90th percentile of standard deviation σ_u
τ	turbulent momentum flux
Δs	fractional speed-up (flow over hills)
Θ	potential temperature

Θ_v	virtual potential temperature
Ω	rotational speed of the Earth
Ψ_m	integral stability correction function
Λ_1	length scale
Λ_u	length scale
Λ_{smax}	spectral length scale
∇	nabla (differential operator)

Chapter 1
Introduction

The available wind energy, E_{wind} in atmospheric flow, i.e. the kinetic energy of the air, 0.5 ρu^2 advected with the wind, u is quantified by the following relation:

$$E_{wind} = 0.5\rho A_r u^2 u = 0.5\rho A_r u^3 \quad (1.1)$$

where ρ is air density, A_r is the rotor area of the turbine and u is the average wind speed over the rotor area. Equation (1.1) gives the available wind energy over the rotor disk in Watt when the air density is given in kg/m^3, the rotor area in m^2 and the wind speed in m/s. Theoretically, turbines can extract up to 16/27 of this amount (Betz 1926). It is an engineering issue how close one can come to this theoretical limit. This is not discussed in this book. The other challenge is that wind speed and air density are not a constant. This book is mainly about how wind speeds vary with space (especially in vertical direction) and time in the atmospheric boundary layer. Air density is addressed in Sect. 2.7. This part of the discipline meteorology is called "wind energy meteorology" today. We will start with some basic thoughts on wind energy and a description of the structure of this book in this introduction before we will start to determine the wind speed and air density and its variations in Chap. 2.

1.1 Scope of the Book

Mankind's need for energy will persist or even increase for the foreseeable future. A sustainable supply will only be possible from renewable energies in the long run. The presently used fossil energies are limited in their resources, produce air pollutants during combustion and endanger the Earth's climate. Renewable energies comprise water power, wave and tidal energy, geothermal energy, biomass, solar energy, and—last but not least—wind energy. This volume focuses on the atmospheric conditions which permit the generation of electricity from wind energy by wind turbines. It has been written from the viewpoint of a meteorologist who has many years of experience with the demands in wind energy generation.

S. Emeis, *Wind Energy Meteorology*, Green Energy and Technology,
DOI: 10.1007/978-3-642-30523-8_1,

Systematic electricity generation from the wind has been performed for more than 20 years. In the early years, the turbines were small, rotor diameters being much smaller than the vertical extent of the atmospheric surface layer. In those times it was relatively easy to assess the local wind climate in order to calculate turbine loads and energy yields. The knowledge of the frequency distribution of the mean wind speed at hub height and the overall turbulence intensity was sufficient to supply the necessary background information for the siting of single turbines and small wind parks.

In the meantime, the size of turbines has increased. The hub height of multi-MW turbines is often above the atmospheric surface layer and rotor diameters of more than 100 m are frequently found. Offshore turbines with diameters of more than 160 m and a power of 7 MW have already been designed and will be deployed in the near future. This leads to much more complicated interactions between the turbines and the lower atmosphere. Meteorological features which had been considered as irrelevant for a long time are now becoming decisive for planning and running single large turbines and increasingly larger wind parks. In particular, vertical gradients in mean wind speed as well as in turbulence intensity have to be known. Furthermore, the vertical range for which these wind parameters must be obtained has now moved to heights which are hardly reachable by masts. New measurement techniques are required to collect the necessary wind information. This has led to a boom in surface-based remote sensing techniques (see Emeis 2011). The economic success of wind turbines depends on a precisely determined trade off between erection and operation costs and wind energy yields. Each additional meter in hub height is only meaningful if the higher yields pay the additional costs.

Additionally, especially in countries adjacent to the North Sea and the Baltic, the main area for wind park development has moved from land to marine sites. Here, offshore wind parks will probably deliver most of the wind energy in the future. This means that wind parks are now erected in areas where many details of the vertical structure of the atmospheric boundary layer are not sufficiently known. Experimental data from the marine boundary layer are available—if any—for only a shallow layer previously explored from buoys, ships, and oil racks. A few masts, like the three German 100 m high FINO masts, have been erected recently in the German Bight and the Baltic. They are presently delivering long-term information on a deeper layer of the marine boundary layer for the first time.

This book tries to analyse and summarize the now existing information of atmospheric boundary layers—onshore and offshore—with respect to wind power generation. The presentation will focus on the vertical profiles of wind and turbulence. It tries to explain the physical processes behind the observable vertical profiles. It will not display wind climatologies for certain regions of the world. The analysis will include features like vertical profile laws beyond those power laws which had been suitable for the surface layer assessment for a long time, instationary phenomena like nocturnal low-level jets, the wind-speed dependent roughness and turbulence conditions in marine boundary layers, and the complex wind-wakes interactions in larger wind parks.

1.2 Overview of Existing Literature

No monograph which is solely devoted to the meteorological basics of wind energy generation is known to the author apart from a WMO Technical note on "Meteorological Aspects of the Utilization of Wind as an Energy Source" which appeared in 1981 and did not anticipate the size of today's turbines. There is a larger body of literature on winds and turbulence in the atmospheric boundary layer appearing in many monographs and journals, but only a smaller number of these papers make reference to wind energy generation (see, e.g., Petersen et al. 1998a, b). On the other hand there are already many books and papers on wind energy generation itself. These existing books mainly concentrate on technical and engineering issues and cover the wind resources in just one or a few chapters. A very recent example is the second edition of the "Wind Energy Handbook" by Burton et al. (2011). Chapter 2 of this book summarizes wind speed variations, gusts and extreme wind speeds, wind speed prediction, and turbulence within 30 pages. Likewise, Hau in his book on "Wind turbines", published by Springer in 2006, summarizes the wind resources in its Chap. 13 in 34 pages. A monograph on the special field of wind speed forecasts is "Physical approach to short-term wind power prediction" by Lange and Focken 2006, which was published by Springer in 2006.

1.3 History of Wind Energy Generation

Mankind has always used the power of the wind for its purposes. This started with the separation of chaff from wheat and other cereals and the air-conditioning of buildings in subtropical and tropical areas. Winds were used to maintain fires and to melt metals. Sailing ships were invented in order to travel over the seas and to establish trade relations with remote coasts. The nearly constantly blowing winds in the subtropical belts of the Earth are still named "trade winds" today.

Wind mills date back at least 2000 years. Heron of Alexandria, who lived in the first century AD, is said to be the first to have invented a wind-driven wheel. His machine was merely used to drive organ pipes (Brockhaus, vol. 24, 2001). Wind mills in Persia are said to have existed from the seventh century AD (Neumann 1907) or from the tenth century (Brockhaus 2001). Those were cereal mills with a vertical axis (Hau 2000). The first wind mill in France is mentioned in 1105 (Neumann 1907). From there, this technology spread into England, where the first ones arose in 1140 (Neumann 1907). They appear in growing numbers in eastern parts of England and Northern Europe in the thirteenth century, e.g., 1235 in Denmark. The climax of this development is found between 1500 and 1650 when the arable surface of the Netherlands could be extended by 40 % due to the use of wind-driven drainage pumps (DeBlieu 2000). The first German wind mill is said to have been erected in Speyer in 1393 (Neumann 1907). About 100,000 wind mills were operated in Europe for the purpose of pumping water and producing flour in the eighteenth and nineteenth century, an era ending however, with the advent of steam engines and electricity. See Ackermann and Söder (2000) for further historical notes.

The history of producing electrical energy from the wind is much shorter. The Dane Poul la Cour (1846–1908) built the first wind turbine in Askov (Denmark) in 1891. The German engineer Betz described the aerodynamic theory of wind turbines for the first time in 1926 (Betz 1926). Betz' factor (16/27) is still known today as a limiting factor for the amount of the energy which can be extracted from wind flow by a turbine.

But it is not before the last two decades of the twentieth century that wind turbines have been erected in larger numbers and growing sizes. An early failed attempt was the construction of the German 3 MW turbine "Growian" (**gro**ße **Wi**ndenergie**an**lage) in 1983. It was a two-blade turbine with a rotor diameter of 100 m. It produced electricity for only 17 days due to a number of technical problems and was removed in 1988. Development was then re-started beginning with small turbines. This "evolutionary" approach was successful so that today even larger turbines than Growian are standard, especially for offshore wind parks.

1.4 Potential of Wind Energy Generation

Wind energy is a renewable form of energy. It is available nearly all over the world, though having considerable regional differences. Wind energy forms from solar energy and is replenished by it continuously. Solar energy is practically available without any limits. The transformation from solar energy into wind energy does not involve the carbon cycle either, with the exception of the production, transport, erection and maintenance of the turbines. Wind energy results from horizontal air pressure differences which in turn are mainly due to latitudinal differences in solar irradiation. In the natural planetary atmospheric energy cycle, wind energy is mostly dissipated by friction occurring mainly at the Earth's surface and is thus transformed into the last and lowest-ranking member of the planetary energy chain: heat. Generation of electrical energy from the wind does not really disturb this planetary energy cycle. It just introduces another near-surface frictional force which partially produces higher-valued electrical energy and only partially heat. When this electrical energy is used by mankind it is also transformed into heat and the planetary energy cycle is closed again. As electrical energy is practically used without any delay and the conservation law for energy is not disturbed, the global planetary energy cycle seems to be undisturbed by energy production from the wind. Therefore, wind power can be considered as a sustainable form of renewable energy. But the entropy budget is affected as well. Large-scale energy production from the wind increases the entropy in the Earth system and could slow down atmospheric circulations. See Sect. 7.4 for further discussions on the interaction between wind power generation and climate.

The globally available energy in the wind can be estimated from the chain of energy conversions in the Earth's atmosphere [the numbers given here are based on earlier seminal publications such as by Lorenz (1955) and Peixoto and Oort (1992)]. The incoming solar power at the top of the atmosphere is roughly 174 300 TW (~342 W/m^2). 1743 TW (~3.5 W/m^2 or 55,000 EJ/year) of this power is available in form of kinetic energy that will eventually be dissipated in the atmosphere. About half of this dissipation takes place in the boundary layer (871 TW or 1.75 W/m^2). This yields 122 TW of potential power assuming that one fourth of the Earth's surface is accessible for wind energy generation and that wind turbines can theoretically extract up to 56 % of this energy (Betz' limit). Practically, maybe 50 % of this is realistic, meaning that the total potential wind power extractability is about 61 TW (1,925 EJ/year). Other estimates which use similar approaches come to energy amounts of the same magnitude (see e.g., Miller et al. (2011) who derive 18–68 TW). A more pessimistic evaluation by de Castro et al. (2011) starts with 1200 TW for the global kinetic energy of the Earth's atmosphere. 8.3 % of this energy is available in a 200 m deep surface layer giving 100 TW. 20 % of the land surface is suitable for the extraction of this surface layer energy giving 20 TW. Restricting wind parks to areas with reasonable wind resources halves this further to 10 TW. Then de Castro et al. estimate that only 10 % of this energy can be extracted by wind turbines. Thus, their estimation is that just 1 TW (32 EJ/year) is the amount of energy extractable from the wind.

While the estimate of the global kinetic energy in the atmosphere is rather robust and yields probably more than 1,000 TW, the two critical assumptions in these calculations are the share of this energy that is dissipated at the surface (here varying between 8 and 50 %) and the share which can be extracted from this near-surface kinetic energy due to technical aspects of the turbines (here varying between 10 and 50 %). Probably a single-digit number given in TW is a realistic estimate for the wind energy available from the Earth's atmosphere.

These numbers have to be compared to the total energy demand of mankind which presently is roughly 15 TW (443 EJ/year) and which is expected to rise to about 30 TW (947 EJ/year) by the middle of the century and 45 TW (1420 EJ/year) by the end of the century (CCSP 2007). This comparison makes clear that wind energy can only be part of the solution for a supply of mankind with renewable energies. Other forms of renewable energies have to be exploited in parallel. Furthermore, it can be expected that energy extractions of even 10 % of the available wind energy will already have considerable effects on the Earth's climate (see Sect. 7.4).

1.5 Present Status of Wind Energy Generation

The worldwide wind energy conversion capacity reached 215 GW by the end of June 2011, out of which 18.4 GW were added in the first 6 months of 2011

(WWEA[1]). The largest share of this has been erected in China (52.8 GW) followed by USA (42.4 GW) and Germany (nearly 28 GW). Spain has an installed capacity of 21.2 GW and India of 14.6 GW. China has more than doubled its capacity since the end of 2009. These 215 GW deliver about 2.5 % of the global energy demand (GWEC Global Wind Energy Outlook 2010[2]). In Europe this share was 5.3 % at the end of 2010 (EWEA)[3] and 9.5 % of the net electric energy consumption in Germany (Ender 2011). Substantial increases in these shares are planned for the next 20 years.

Offshore wind energy production is still in its infancy although gigantic plans for this have been developed. In Germany, 0.21 GW have been installed at the end of June 2011 (Ender 2011), which is less than one percent of the total installed capacity.

The globally installed capacity of 215 GW is already a considerable fraction of the available wind energy of a few TW. The present growth rate of this installed capacity by extrapolating the numbers for the first half of 2011 gives roughly 15 % per year. This rate would lead to a doubling within 6 years and to a tenfold value in nearly 18 years. A steady increase of the installed capacity with this rate of 15 % per year would meet the estimated limits in Sect. 1.4 in about 20–30 years. Thus, it cannot be expected that the present growth rate will prevail for a longer time.

Therefore, the available wind energy should be extracted in a most efficient way. Understanding the meteorological basics for the extraction of wind energy gathered in this book shall help to reach this efficiency.

1.6 Structure of This Book

This publication is organized as follows. Chapter 2 explains the origin of the large-scale winds in our atmosphere and presents the main laws driving atmospheric motion in the free atmosphere. Additionally, the determination of air density is addressed. Chaps. 3–5 present the vertical profiles of wind and turbulence over different surface types. Chapter 3 reviews classical boundary layer meteorology over flat natural homogeneous land surfaces. Emphasis is laid on the vertical extension of wind profiles from the surface layer into the Ekman layer above, since large multi-MW wind turbines reach well into this layer today. This includes the description of nocturnal low-level jets, which lead to nocturnal maxima in wind energy conversion with large turbines. Internal boundary layers forming at step changes of the surface properties, forest boundary layers and urban boundary layers are shortly addressed at

[1] http://www.indea.org/home/index.php?option=comcontent&task=view&id=317&Itemid=43 (read Dec 14 2011)

[2] http://www.gwec.net/fileadmin/documents/Publications/GWEO%202010%20final.pdf (read Dec14 2011)

[3] http://www.ewea.org/fileadmin/ewea_documents/documents/statistics/EWEA_Annual_Statistics_2010.pdf (read Dec 14, 2011)

the end of this chapter. Chapter 4 highlights the peculiarities of flow over complex terrain, especially of orography. Basic features such as speed-up over hills are derived using a simple analytical model. A separate description of flow over this surface type is relevant, because the near-coastal flat areas are often sufficiently used today and sites more inland have to be analysed for future wind energy production. The deployment of turbines far away from the coasts closer to urban and industrial areas also helps to reduce the erection of massive power lines connecting generation and consumption areas. The last of these three chapters on vertical profiles, Chap. 5 deals with a surface type which presently is becoming more and more important: the marine boundary layer over the sea surface. The planning of huge offshore wind parks require that considerable space is devoted to this surface type. Chapter 6 looks into the features and problems which come with large wind parks over any of the aforementioned surface types. This is no longer a pure meteorological topic, because the properties of the wind turbines and their spatial arrangement in the park become important as well. This chapter will present another simple analytical model which can be used to make first estimates on the influence of surface roughness and thermal stability of the atmosphere as well as the influence of the turbines' thrust coefficient and the mean distance of the turbines within the wind park on the overall efficiency of the wind park.

Chapters 3–6 all end with a short summary on the main aspects which should be taken into account from a meteorological point of view when planning and running wind turbines. Chapter 7 gives an outlook on possible future developments and certain limitations to large-scale wind energy conversion. Appendix A summarizes the different parameters which are frequently used to describe the properties of the wind. Here, the distinction between mean winds and turbulent motion is introduced and basic statistical concepts are described. Appendix B introduces into techniques to determine the mixed layer height—an input parameter in the description of wind profiles extending above the surface layer—from surface-based remote sensing. Surface-based remote sensing has today become a major tool to probe the conditions of the atmospheric boundary layer.

References

Ackermann, T., L. Söder: Wind energyWind energy technology and current status: a review. Renew. Sustain. Energy Rev. 4, 315–374 (2000)

Betz, A.: Wind-Energie und ihre Ausnutzung durch Windmühlen. Vandenhoeck & Ruprecht, Göttingen. 64 S. (1926)

Brockhaus, F.A. Brockhaus Enzyklopädie. Vol. 24. Gütersloh (2001)

Burton, T., N. Jenkins, D. Sharpe, E. Bossanyi: Wind Energy Handbook. Second edition, John Wiley & Sons, 742 pp. (2011)

Castro, de C., M. Mediavilla, L.J. Miguel, F. Frechoso: Global Wind Power Potential: Physical and Technological Limits. Energy Policy 39, 6677–6682 (2011)

CCSP: US Climate Change Science Program. Synthesis and Assessment Product 2.1a, July 2007 (2007). http://www.climatescience.gov/Library/sap/sap2-1/finalreport/sap2-1a-final-all.pdf (read Dec 17, 2011)

DeBlieu, J.: Vom Wind. Wie die Luftströme Leben, Land und Leute prägen. Wilhelm Goldmann Verlag, München (btb Taschenbuch 72611). 411 pp. (2000)
Emeis, S.: Surface-Based Remote Sensing of the Atmospheric Boundary Layer. Series: Atmospheric and Oceanographic Sciences Library, Vol. 40. Springer Heidelberg etc., X + 174 pp. (2011)
Ender, C.: Wind Energy Use in Germany – Status 30.06.2011. DEWI Mag. **39**, 40-49 (2011)
Hau, E.: Wind Turbines. Springer Berlin. 624 pp. (2000)
Lange, M., U. Focken: Physical Approach to Short-Term Wind Power Prediction. Springer Berlin. 208 pp. (2006)
Lorenz, E.: Available potential energy and the maintenance of the general circulation. Tellus 7, 271-281 (1955)
Miller, L.M., F. Gans, A. Kleidon: Estimating maximum global land surface wind power extractability and associated climatic consequences. Earth Syst. Dynam. 2, 1-12 (2011)
Neumann, F.: Die Windkraftmaschinen. Voigt Leipzig. 174 S. (1907)
Peixoto, J.P., A.H. Oort: Physics of Climate. Springer Berlin etc., 520 pp. (1992)
Petersen, E.L., N.G. Mortensen, L. Landberg, J. Højstrup, H.P. Frank: Wind power meteorology. Part I. Climate and turbulence. Wind Energy, 1, 25-45 (1998a)
Petersen, E.L., N.G. Mortensen, L. Landberg, J. Højstrup, H.P. Frank: Wind Power Meteorology. Part II: Siting and Models. Wind Energy, 1, 55-72 (1998b)

Chapter 2
Wind Regimes

The principle origin of the winds in the Earth's atmosphere and the potentially available power from these winds have been qualitatively described in Sect. 1.4. This general description of the driving forces for the wind has to be brought into a mathematical formulation for precise turbine load and energy yield calculations and predictions. Therefore, this chapter will present the basic wind laws in the free atmosphere. Vertical wind profiles in atmospheric boundary layers over different surface types will be presented in the subsequent Chaps. 3–5.

2.1 Global Circulation

Flow patterns and winds emerge from horizontal surface and atmospheric temperature contrasts on all spatial scales from global to local size. Globally, the tropical belt and the lower latitudes of the Earth are the main input region for solar energy, while the higher latitudes and the poles are the regions with a negative energy balance, i.e. the Earth here loses energy through thermal radiation. Ocean currents and atmospheric heat conduction are not sufficient to compensate for this differential heatingof the globe. The global atmospheric circulation has to take over as well. Main features of this global atmospheric circulation are the Hadley cell, the Ferrel cell and the polar cell which become visible from a latitude-height plot showing an average over all longitudes of the winds in the troposphere and stratosphere. The Hadley cell exhibits a direct thermal circulation. Warm air rises near the equator, moves towards the poles aloft and descends in the subtropics. The region of sinking motion is characterised by large anticyclones in the surface pressure field and deserts. Likewise, the polar cell exhibits a direct thermal circulation as well. Here, cold air sinks over the poles and rises at higher latitudes. This is the reason for generally high pressure over the poles. In between the Hadley cell and the polar cell lies the thermally indirect Ferrel cell. This cell is characterised by rising colder air at higher latitudes and sinking warmer air in the

S. Emeis, *Wind Energy Meteorology*, Green Energy and Technology,
DOI: 10.1007/978-3-642-30523-8_2, © Springer-Verlag Berlin Heidelberg 2013

subtropics. This circulation is indirect and it is the result of the integral effect over all the moving cyclones in this belt of temperate latitudes. Effectively, the Ferrel cells transports warmer air towards the poles near the ground and colder air towards the tropics aloft. This indirect circulation is maintained by energy conversions from potential energy into kinetic energy in the moving cyclones of the temperate latitudes.

The just described system of cells would only produce meridional winds, i.e. winds from South to North or vice versa. The Earth's rotation is modifying this meridional circulation system by the Coriolis force. Winds towards the poles get a westerly component, winds towards the equator an easterly component. Therefore, we mainly observe westerly winds at the ground in the Ferrel cellwhile we observe easterly winds at the ground in the Hadley cell and the polar cell. The north easterly winds near the ground of the Hadley cell are also known as the trade winds. These global wind cells have a spatial scale of roughly 10,000 km. The global wind system is modified by the temperature contrasts between the continents and the surrounding oceans and by large north-south orientated mountain ranges, in particular those at the west coasts of the Americas. These modifications have a spatial scale of some 1,000 km. Even smaller land-sea wind systems in coastal areas may have an order of 100 km; mountain and valley wind systems can be even smaller in the order of several tens of kilometres. All these wind systems may be suitable for wind power generation.

While the trade winds and the winds in the polar cell exhibit quite some regularity and mainly have seasonal variations, the winds in the Ferrel cell are much more variable in space and time. Near-surface wind speeds in normal cyclones can vary between calms and about 25 m/s within a few hours. Wind speeds in strong hibernal storms of the temperate latitudes can reach about 35–40 m/s while wind speeds in subtropical hurricanes easily reach more than 50 m/s. Cut-off wind speeds of modern wind energy turbines are between 25 and 30 m/s. Thus strong storms in temperate latitudes may lead to phases where the wind potential can no longer be used. These hibernal storms are most likely in Northwestern Europe, Northeastern Canada, the Pacific coasts of Canada and Alaska as well as the southern tips of South America, Africa and Australia.

Hurricanes are called typhoons in Southeast Asia and cyclones in India. The occurrence of hurricanes can even threaten the stability of the construction of the turbines, because they can come with wind speeds above those listed in the IEC design standards. The hurricane risks have been investigated by Rose et al. (2012). In particular, the planning of offshore wind parks in hurricane-threatened areas needs special attention. According to the map of natural hazards published by the reassurance company Munich Re, hurricane-prone areas are the southern parts of the Pacific coasts and the Atlantic coasts of the United States and Central America, Eastern India and Southeast Asia, Madagascar and the northern half of Australia.

There are very strong winds on even smaller scales such as thunderstorm downbursts, whirlwinds and tornados, but their variability and destructive force is not suited for windpower generation. Rather turbines have to be constructed in a way that they can stand these destructive forces while being shut off. See also Sects. 2.6 and 6.5 for wind hazards.

2.2 Driving Forces

The equations in the following Subchapters describe the origin and the magnitude of horizontal winds in the atmosphere. We will start with the full set of basic equations in Sects. 2.2.1 and 2.2.2 and will then introduce the usual simplifications which lead to the description of geostrophic and gradient winds in Sect. 2.3. Geostrophic and gradient winds, which blow in the free atmosphere above the atmospheric boundary layer, have to be considered as the relevant external driving force in any wind potential assessment and any load assessment. Vertical variations in the geostrophic and gradient winds are described by the thermal winds introduced in Sect. 2.4.

2.2.1 Hydrostatic Equation

The most basic explanation of the wind involves horizontal heat gradients. The sun heats the Earth's surface differently according to latitude, season and surface properties. This heat is transported upward from the surface into the atmosphere mainly by turbulent sensible and latent heat fluxes. This leads to horizontal temperature gradients in the atmosphere. The density of air, and with this density the vertical distance between two given levels of constant pressure, depends on air temperature. A warmer air mass is less dense and has a larger vertical distance between two given pressure surfaces than a colder air mass. Air pressure is closely related to air density. Air pressure is a measure for the air mass above a given location. Air pressure decreases with height. In the absence of strong vertical accelerations, the following hydrostatic equation describes this decrease:

$$\frac{\partial p}{\partial z} = -g\rho = -\frac{gp}{RT} \tag{2.1}$$

where p is air pressure, z is the vertical coordinate, g is the Earth's gravity, ρ is air density, R is the specific gas constant of air, and T is absolute air temperature. With typical near-surface conditions ($T = 293$ K, $R = 287$ J kg^{-1} K^{-1}, $p = 1{,}000$ hPa and $g = 9.81$ ms^{-2}) air pressure decreases vertically by 1 hPa each 8.6 m. In wintry conditions, when $T = 263$ K, pressure decreases 1 hPa each 7.7 m near the surface. At greater heights, this decrease is smaller because air density is

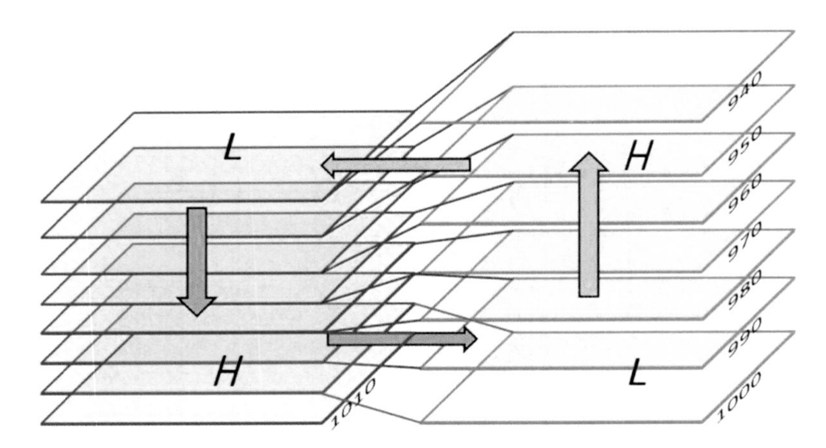

Fig. 2.1 Vertical pressure gradients in warmer (*right*) and colder (*left*) air. Planes symbolizes constant pressure levels. Numbers give air pressure in hPa. Capital letters indicate high (*H*) and low (*L*) pressure at the surface (*lower letters*) and on constant height surfaces aloft (*upper letters*). *Arrows* indicate a thermally direct circulation

decreasing with height as well. At a height of 5.5 km the air pressure is at about half of the surface value, and thus, the pressure only decreases by 1 hPa every 15 m. An (unrealistic) atmosphere at constant near-surface density would only be 8 km high!

The consequence of (2.1) is that the pressure in warm air masses decreases more slowly with height than in cold air masses. Assuming a constant surface pressure, this would result in horizontal pressure gradients aloft. A difference in 30° in air mass temperature will cause a 1.36 hPa pressure gradient between the warm and the cold air mass 100 m above ground. This pressure gradient produces compensating winds which tend to remove these gradients. In reality, surface pressure sinks in the warmer region ("heat low"). This situation is depicted in Fig. 2.1. In a situation with no other acting forces (especially no Coriolis forces due to the rotating Earth) this leads to winds blowing from higher towards lower pressure. Such purely pressure-driven winds are found in land-sea and mountain-valley wind systems. This basic effect is depicted in term III in the momentum budget equations that will be introduced in the following section.

2.2.2 Momentum Budget Equations for the Wind

A mathematical description of the winds is most easily done by considering the momentum balance of the atmosphere. Momentum is mass times velocity. The momentum budget equations are a set of differential equations describing the

Table 2.1 Latitude-dependent Coriolis parameter f in s^{-1} for the northern hemisphere. The values in both columns are negative for the southern hemisphere

Latitude (in degrees)	Coriolis parameter in s^{-1}
30	0.727×10^{-4}
40	0.935×10^{-4}
50	1.114×10^{-4}
60	1.260×10^{-4}

acceleration of the three wind components. In complete mass-specific form, they read (mass-specific means that these equations are formulated per unit mass, the mass-specific momentum has the physical dimension of a velocity. Therefore, we say wind instead of momentum in the following):

$$\frac{\partial u}{\partial t} + \vec{v}\nabla u + \frac{1}{\rho}\frac{\partial p}{\partial x} \qquad - fv + f^{*}w \qquad \mp v\frac{|\vec{v}|}{r} \qquad + F_x = 0 \quad (2.2)$$

$$\frac{\partial v}{\partial t} + \vec{v}\nabla v + \frac{1}{\rho}\frac{\partial p}{\partial y} \qquad + fu \qquad \pm u\frac{|\vec{v}|}{r} \qquad + F_y = 0 \quad (2.3)$$

$$\frac{\partial w}{\partial t} + \vec{v}\nabla w + \frac{1}{\rho}\frac{\partial p}{\partial z} \quad - g \quad - f^{*}u \qquad + F_z = 0 \quad (2.4)$$

I II III IV V VI VII

where u is the wind component blowing into positive x direction (positive in eastward direction), v is the component into y direction (positive in northward direction) and w is the vertical wind (positive upward). The wind vector is $\vec{v} = (u, v, w)$, the horizontal Coriolis parameter is $f = 2\Omega \sin\varphi$ where Ω is the rotational speed of the Earth and φ is the latitude (see Table 2.1), the vertical Coriolis parameter is $f^{*} = 2\Omega \cos\varphi$, r is the radius of curvature, and F_x, F_y, and F_z are the three components of the frictional forces, which will be specified later. The Eqs. (2.2)–(2.4), which are called Eulerian equations of motion in meteorology, are a special form of the Navier-Stokes equations in hydrodynamics.

Term I in Eqs. (2.2)–(2.4) is called inertial or storage term, it describes the temporal variation of the wind components. The non-linear term II expresses the interaction between the three wind components. Term III specifies the above-mentioned pressure force. Term IV, which is present in (2.4) only, gives the influence of the Earth's gravitation. Term V denotes the Coriolis force due to the rotating Earth. Term VI describes the centrifugal force in non-straight movements around pressure maxima and minima (the upper sign is valid for flows around lows, the lower sign for flows around high pressure systems). The last term VII symbolizes the frictional forces due to the turbulent viscosity of air and surface friction.

The terms in (2.2)–(2.4) may have different magnitudes in different weather situations and a scale analysis for a given type of motion may lead to discarding some of them. Nearly always, the terms containing f^* are discarded because they are very small compared to all other terms in the same equation. In larger-scale motions term VI is always neglected as well. Term VI is only important in whirl winds and close to the centre of high and low pressure systems. Looking at the vertical acceleration only (Eq. (2.4)), terms III and IV are dominating. Equating these two terms in (2.4) leads to the hydrostatic equation (2.1) above.

There is only one driving force in Eqs. (2.2–2.4): the abovementioned pressure force which is expressed by term III. The constant outer force due to the gravity of the Earth (term IV) prevents the atmosphere from escaping into space. The only braking force is the frictional force in term VII. The other terms (II, V, and VI) just redistribute the momentum between the three different wind components. Thus, sometimes terms V and VI are named "apparent forces". In the special case when all terms II to VII would disappear simultaneously or would cancel each other perfectly, the air would move inertially at constant speed. This is the reason why term I is often called inertial term.

2.3 Geostrophic Winds and Gradient Winds

The easiest and most fundamental balance of forces is found in the free troposphere above the atmospheric boundary layer, because frictional forces are negligible there. Therefore, our analysis is started here for large-scale winds in the free troposphere. The frictional forces in term VII in Eqs. (2.2–2.4) can be neglected above the atmospheric boundary layer. Term VI is also very small and negligible away from pressure maxima and minima. The same applies to term II for large-scale motions with small horizontal gradients in the wind field. A scale analysis shows that the equilibrium of pressure and Coriolis forcess is the dominating feature and the inertial term I can be neglected as well. This leads to the following two equations:

$$-\rho f u_g = \frac{\partial p}{\partial y} \tag{2.5}$$

$$\rho f v_g = \frac{\partial p}{\partial x} \tag{2.6}$$

with u_g and v_g being the components of this equilibrium wind, which is usually called geostrophic wind in meteorology. The geostrophic wind is solely determined by the large-scale horizontal pressure gradient and the latitude-dependent Coriolis parameter, the latter being in the order of 0.0001 s^{-1} (see Table 2.1 for some sample values). Because term VII had been neglected in the definition of the geostrophic wind, surface friction and the atmospheric stability of the atmospheric boundary layer has no influence on the magnitude and direction of the geostrophic wind. The modulus of the geostrophic wind reads:

$$|v_g| = \sqrt{u_g^2 + v_g^2} \tag{2.7}$$

The geostrophic wind blows parallel to the isobars of the pressure field on constant height surfaces. Following Eqs. (2.5) and (2.6), a horizontal pressure gradient of about 1 hPa per 1,000 km leads to a geostrophic wind speed of about 1 m/s. In the northern hemisphere, the geostrophic wind blows counter-clockwise around low pressure systems and clockwise around high pressure systems. In the southern hemisphere the sense of rotation is opposite.

Term VI in Eqs. (2.2–2.4) is not negligible in case of considerably curved isobars. The equilibrium wind is the so-called gradient wind in this case:

$$-\rho f u = \frac{\partial p}{\partial y} \pm \frac{\rho u|\vec{v}|}{r} \tag{2.8}$$

$$\rho f v = \frac{\partial p}{\partial x} \mp \frac{\rho v|\vec{v}|}{r} \tag{2.9}$$

Once again, the upper sign is valid for flows around lows, the lower sign for flows around high pressure systems. The gradient wind around low pressure systems is a bit lower than the geostrophic wind (because centrifugal force and pressure gradient force are opposite to each other), while the gradient wind around high pressure systems is a bit higher than the geostrophic wind (here centrifugal force and pressure gradient force are unidirectional).

Sometimes, in rare occasions, the curvature of the isobars can be so strong that the centrifugal force in term VI is much larger than the Coriolis force in term V so that an equilibrium wind forms which is governed by pressure forces and centrifugal forces only. This wind, called cyclostrophic wind by meteorologists, is found in whirl winds and tornados.

The geostrophic wind and the gradient wind are not height-independent in reality. Horizontal temperature gradients on levels of constant pressure lead to vertical gradients in these winds. The wind difference between the geostrophic winds or gradient winds at two different heights is called the thermal wind.

2.4 Thermal Winds

We introduced in Sect. 2.3 the geostrophic wind as the simplest choice for the governing large-scale forcing of the near-surface wind field. The geostrophic wind is an idealized wind which originates from the equilibrium between pressure gradient force and Coriolis force. Until now we have always anticipated a barotropic atmosphere within which the geostrophic wind is independent of height, because we assumed that the horizontal pressure gradients in term III of (2.2) and (2.3) are independent of height. This is not necessarily true in reality and the deviation from a height-independent geostrophic wind can give an additional contribution to the vertical wind profile as well. The horizontal pressure gradient becomes height-dependent in an atmosphere with a large-scale horizontal temperature gradient. Such an atmosphere is called baroclinic and the difference in the wind vector between geostrophic winds at two heights is called thermal wind. The real atmosphere is nearly always at least slightly baroclinic, thus the thermal wind is a general phenomenon.

Thermal winds do not depend on surface properties. So they can appear over all surface types addressed in Chaps. 3–5.

Differentiation of the hydrostatic equation (2.1) with respect to y and differentiation of the definition equation for the u-component of the geostrophic wind (2.5) with respect to z leads after the introduction of a vertically averaged temperature T_M to the following relation for the height change of the west–east wind component u:

$$\frac{\partial u}{\partial z} = -\frac{g}{fT_M}\frac{\partial T_M}{\partial y} \tag{2.10}$$

Subsequent integration over the vertical coordinate from the roughness length z_0 to a height z gives finally for the west–east wind component at the height z:

$$u(z) = u(z_0) - \frac{g(z - z_0)}{fT_M}\frac{\partial T_M}{\partial y} \tag{2.11}$$

The difference between $u(z)$ and $u(z_0)$ is the u-component of the thermal wind. A similar equation can be derived for the south–north wind component v from Eqs. (2.1) and (2.6):

$$v(z) = v(z_0) + \frac{g(z - z_0)}{fT_M}\frac{\partial T_M}{\partial x} \tag{2.12}$$

Following (2.10) and (2.11), the increase of the west–east wind component with height is proportional to the south–north decrease of the vertically averaged temperature in the layer between z_0 and z. Likewise, (2.12) tells us that the south–north wind component increases with height under the influence of a west–east temperature increase. Usually, we have falling temperatures when travelling north in the west wind belt of the temperate latitudes on the northern hemisphere, so we usually have a vertically increasing west wind on the northern hemisphere.

Equations (2.11) and (2.12) allow for an estimation of the magnitude of the vertical shear of the geostrophic wind, i.e. the thermal wind from the large-scale horizontal temperature gradient. The constant factor $g/(fT_M)$ is about 350 m/(s K). Therefore, a quite realistic south–north temperature gradient of 10^{-5} K/m (i.e., 10 K per 1,000 km) leads to a non-negligible vertical increase of the west–east wind component of 0.35 m/s per 100 m height difference.

The thermal wind also gives the explanation for the vertically turning winds during episodes of cold air or warm air advection. Imagine a west wind blowing from a colder to a warmer region. Equation (2.12) then gives an increase in the south–north wind component with height in this situation. This leads to a backing of the wind with height. In the opposite case of warm air advection the wind veers with height.

2.5 Boundary Layer Winds

The wind speed in the atmospheric boundary layer must decrease to zero towards the surface due to the surface friction (no-slip condition). The atmospheric boundary layer can principally be divided into three layers in the vertical. The lowest layer which is only a few millimetres deep is laminar and of no relevance for wind energy applications. Then follows the surface layer (also called constant-flux layer or Prandtl layer), which may be up to about 100 m deep, where the forces due to the turbulent viscosity of the air dominate, and within which the wind speed increases strongly with height. The third and upper layer, which usually covers 90 % of the boundary layer, is the Ekman layer. Here, the rotational Coriolis force is important and causes a turning of the wind direction with height. The depth of the boundary layer usually varies between about 100 m at night with low winds and about 2–3 km at daytime with strong solar irradiance.

Scale analysis of the momentum Eqs. (2.2–2.4) for the boundary layer show the dominance of terms III, V, and VII. Sometimes, for low winds in small-scale motions and near the equator, the pressure force (term III) is the only force and a so-called Euler wind develops, which blows from higher pressure towards lower pressure. Such nearly frictionless flows rarely appear in reality. Usually an equilibrium between the pressure force and the frictional forces (terms III and VII) is observed in the Prandtl layer, and an equilibrium between the pressure force, the Coriolis force and the frictional forces (terms III, V, and VII) is observed in the Ekman layer. The Prandtl layer wind is sometimes called antitriptic wind. No equation for the antitriptic winds analog to (2.5, 2.6) or (2.8, 2.9) is available, since neither term III nor term VII contains explicitly the wind speed.

The Prandtl layer is characterised by vertical wind gradients. The discussion of Prandtl layer wind laws which describe these vertical wind speed gradients is postponed to Chap. 3. The vertical gradients are much smaller in the Ekman layer, so that it is meaningful to look at two special cases of (2.2) and (2.3) in the following section.

In a stationary Ekman layer the terms III, V, and VII balance each other, because term I vanishes. This layer is named from the Swedish physicist and oceanographer W. Ekman (1874–1954), who for the first time derived mathematically the influence of the Earth's rotation on marine and atmospheric flows. A prominent wind feature in the Ekman layer is the turning of wind direction with height.

The vertical profiles of these boundary layer winds over different surface types will be analysed in more detail in the upcoming Chaps. 3–5.

2.6 Thunderstorm Gusts and Tornados

There are strong winds which cannot be used for wind energy generation, because they are short-lived and rare in time and place, such that their occurrence is nearly unpredictable. Most prominent among these phenomena are thunderstorm gusts and tornadoes. Offshore tornadoes are called waterspouts. They can be so violent that they can damage wind turbines. Therefore, the probability of their occurrence and their possible strength should be nevertheless investigated during the procedure of wind turbine sitting.

While onshore tornadoes mostly form in the afternoon and the early evening at cold fronts or with large thunderstorms when surface heating is at a maximum, offshore waterspouts are more frequent in the morning and around noon when the instability of the marine boundary layer is strongest due to nearly constant sea surface temperatures (SST) and cooling of the air aloft overnight (Dotzek et al. 2010). However, the seasonal cycle is different. Onshore tornadoes most frequently occur in late spring and summer. Offshore waterspouts peak in late summer and early autumn. In this season, the sea surface temperature of shallow coastal waters is still high, while the first autumnal rushes of cold air from the polar regions can lead to an unstable marine boundary layer favourable for waterspout formation (Dotzek et al. 2010).

Although the characteristics of tornado formation are understood in principle today, the prediction of their actual occurrence remains difficult because a variety of different favourable conditions have to be met simultaneously. In general, following Houze (1993) and Doswell (2001), tornado formation depends largely on the following conditions:

- (potential) instability with dry and cold air masses above a boundary layer capped by a stable layer preventing premature release of the instability;
- a high level of moisture in the boundary layer leading to low cloud bases;
- strong vertical wind shear (in particular for mesocyclonic thunderstorms);
- pre-existing boundary layer vertical vorticity (in particular for non-mesocyclonic convection).

A rough estimation how often a tornado could hit a large wind park is given in Sect. 6.5.

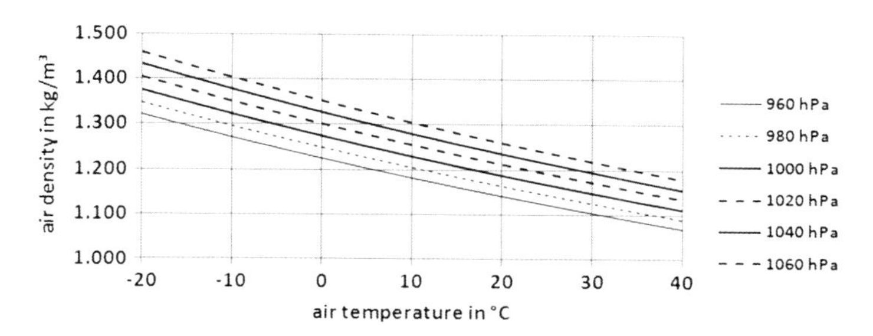

Fig. 2.2 Near-surface air density as function of air temperature and surface pressure

2.7 Air Density

Apart from wind speed, the kinetic energy content of the atmosphere also depends linearly on air density (see Eq. (1.1)). Near-surface air density, ρ is a direct function of atmospheric surface pressure, p and an inverse function of air temperature, T. We have from the state equation for ideal gases:

$$\rho = \frac{p}{RT} \tag{2.13}$$

where $R = 287 \text{ J kg}^{-1} \text{ K}^{-1}$ is the universal gas constant. Equation (2.13) is equivalent to the hydrostatic equation (2.1) above. Figure 2.2 shows air density for commonly occurring values of surface temperature and surface pressure. The Figure illustrates that air density can be quite variable. A cold wintertime high pressure situation could easily come with a density around 1.4 kg/m^3, while a warm low pressure situation exhibits an air density of about 1.15 kg/m^3. This is a difference in the order of 20 %.

Figure 2.2 is valid for a dry atmosphere. Usually the atmosphere is not completely dry and the modifying effect of atmospheric humidity has to be considered. Humid air is less dense than completely dry air. Meteorologists have invented the definition of an artificial temperature which is called virtual temperature. The virtual temperature, T_v is the temperature which a completely dry air mass must have in order to have the same density as the humid air at the actual temperature, T. The virtual temperature is defined as:

$$T_v = T(1 + 0.609q) \tag{2.14}$$

where q is the specific humidity of the air mass given in kg of water vapour per kg of moist air. The temperatures in Eq. (2.14) must be given in K. The difference between the actual and the virtual temperature is small for cold air masses and low specific humidity, but can be several degrees for warm and very humid air masses. Figure 2.2 can be used to estimate air density of humid air masses, if the

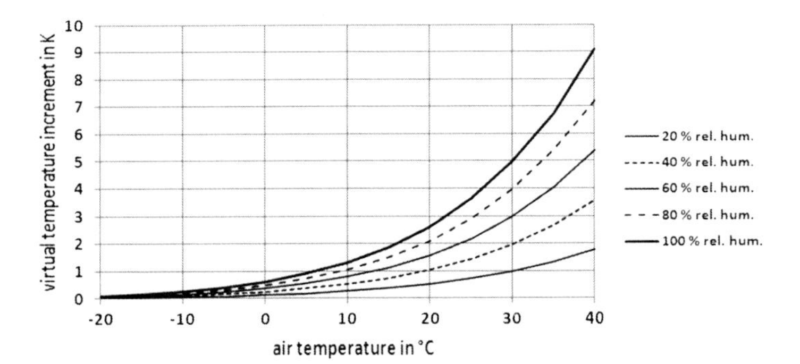

Fig. 2.3 Virtual temperature increment *Tv–T* in K as function of air temperature and relative humidity for an air pressure of 1,013.25 hPa

temperature in Fig. 2.2 is replaced by the virtual temperature. Figure 2.3 gives the increment $T_v - T$ by which the virtual temperature is higher than the actual air temperature as function of temperature and relative humidity of the air for an air pressure of 1,013.25 hPa.

Figure 2.3 shows that the virtual temperature increment is always less than 1 K for temperatures below the freezing point, but reaches, e.g. 5 K for saturated humid air at 30 °C. The virtual temperature increment slightly decreases with increasing air pressure. A 1 % increase in air pressure (10 hPa) leads to a 1 % decrease in the virtual temperature increment. Thus, the determination of the exact density of an air mass requires the measurement of air pressure, air temperature and humidity.

Air density decreases with height, because air pressure decreases with height as given in (2.1). We get from (2.1) to (2.13) (Ackermann and Söder 2000):

$$\rho(z) = \frac{p_r}{R\overline{T}} \exp\left(\frac{-g(z - z_r)}{R\overline{T}}\right) \tag{2.15}$$

p_r is the air pressure at a reference level z_r and $\overline{T}$ is the vertical mean temperature of the layer over which the density decrease is computed. Temperature is decreasing with height as well; therefore equation (2.15) should only be used for small vertical intervals.

References

Ackermann, T., L. Söder: Wind energy technology and current status: a review. Renew. Sustain. Energy Rev. 4, 315–374 (2000)

Doswell, C. A., (Ed.): Severe Convective Storms. Meteor. Monogr. **28**(50), 561 pp. (2001)

Dotzek, N., S. Emeis, C. Lefebvre, J. Gerpott: Waterspouts over the North and Baltic Seas: Observations and climatology, prediction and reporting. Meteorol. Z. 19, 115–129 (2010)

Houze, R.A.: Cloud Dynamics. Academic Press, San Diego, 570 pp. (1993)

Rose, S., P. Jaramillo, M.J. Small, I. Grossmann, J. Apt: Quantifying the hurricane risk to offshore wind turbines. PNAS, published ahead of print February 13, 2012, doi:10.1073/pnas.1111769109 (2012)2

Chapter 3
Vertical Profiles Over Flat Terrain

This chapter is going to introduce the basic laws for the shape of the vertical profiles of wind speed and turbulence in a flat, horizontally homogeneous atmospheric boundary layer (ABL) over land, because this is the simplest surface type. See Chap. 4 for orographically structured complex terrain and Chap. 5 for the marine ABL. The ABL is the lower part of the troposphere and by this the lowest layer of the atmosphere as a whole. In contrast to the free atmosphere above, which had been introduced in Chap. 2, the presence of the underlying Earth's surface has a measurable influence on the ABL. It is the only part of the atmosphere where frictional forces play an important role, and where the temperature and atmospheric stability can exhibit diurnal and annual variations. The ABL acts as a kind of broker that communicates the transport of energy, momentum, and other substances between the Earth surface and the free atmosphere, i.e. the ABL is dominated by vertical fluxes of these variables. These fluxes have their largest amounts directly at the surface and vanish at the top of the ABL. We will concentrate here on vertical wind and turbulence profiles, because these are the most important ABL feature for the generation of energy from the wind. More general descriptions of the ABL can be found in Stull (1988), Arya (1995), Garratt (1992) and other books. Because the ABL enwraps the whole Earth, it is often also called planetary boundary layer (PBL).

The wind speed profile laws for the ABL form the basis for vertical interpolation and/or extrapolation from measurement or model layer heights to hub height or other heights in the rotor plane of a wind turbine. The profile laws also indicate the vertical wind shear which has to be expected across the rotor plane of a turbine. The growing hub heights of modern wind turbines require a careful investigation of the vertical structure of the boundary layer in order to describe the wind profiles correctly. Hub heights of 80 m and more are usually above the surface layer which forms about one tenth of the depth of the total boundary layer. Simple power law or logarithmic profiles are strictly valid in the surface layer only.

The Earth's surface is a place where turbulence is generated, it is usually a sink for atmospheric momentum, and either a source or a sink for heat and moisture. Therefore, we find in the ABL less momentum but more turbulence, and different

S. Emeis, *Wind Energy Meteorology*, Green Energy and Technology,
DOI: 10.1007/978-3-642-30523-8_3, © Springer-Verlag Berlin Heidelberg 2013

heat and moisture concentrations than in the free atmosphere above. The detection of the vertical profiles of the just mentioned atmospheric variables can thus help to identify the vertical structure and extent of the ABL.

Three principal types of the ABL can be distinguished: (1) if heat input from below dominates we find a convective boundary layer (CBL), (2) if the atmosphere is cooled from below we find a stable boundary layer (SBL), and (3) if the heat flux at the lower surface is vanishing and dynamical shear forces are dominating, we find a neutral or dynamical boundary layer. We will start with the principal description of the vertical structure of these three ABL types in the section below.

The vertical structure of these three ABL types additionally depends to a large extent on the type and texture of the underlying surface. Its shape, roughness, albedo, moisture content, heat emissivity, and heat capacity determine the momentum and energy exchange between the surface and the atmosphere. The vertical extent of the ABL is mainly determined by the generation of turbulent kinetic energy at and the input of heat from the lower surface. The following chapters and sections will present some of the most important characteristics of the ABL with respect to the surface features as found, e.g., within the urban boundary layer (UBL, Sect. 3.7) or the marine boundary layer (MBL, Chap. 5). In theory, these characteristics will only appear if the flow is in equilibrium with the underlying surface. Each time when the horizontal atmospheric flow crosses a boundary from one surface type or subtype to the next a new internal boundary layer forms which will eventually—if no further change in surface conditions takes place—reach a new equilibrium. Wind profiles within internal boundary layers are presented in Sect. 3.5.

The simplest structure of the ABL is found over flat, horizontally homogeneous terrain with uniform soil type and land use and a uniform distribution of roughness elements. Its vertical stratification in the roughness sublayer, constant-flux sublayer (Prandtl layer) and Ekman layer is depicted in Fig. 3.1. The evolution of the flat-terrain ABL is mainly determined by the diurnal variation of the energy balance of the Earth's surface. During daytime when the sun is heating the ground, a convective boundary layer (CBL) is growing due to the input of heat from below which generates thermal convection. The CBL is dominated by intense vertical mixing and thus small vertical gradients. During night-time when the ground cools due to the emission of long-wave radiation, a new nocturnal SBL forms near the ground (see Fig. 3.2). The SBL is characterized by low turbulence intensity and large vertical gradients. If clouds, wind, and precipitation override the influence of short-wave and long-wave radiation, the ABL is even simpler and a neutral boundary layer with nearly no diurnal variation forms. Its depth is then mainly determined by the magnitude of the wind shear within it and by the advection of warmer or colder air masses aloft with their own prescribed thermal stratification.

Apart from a viscous or laminar sublayer directly above the surface that is only a few millimetres deep (too shallow in order to be shown in Fig. 3.1), we have two main compartments of the ABL which must be distinguished by the balance of forces within them: (1) the surface (Prandtl) layer or constant-flux layer and (2) the Ekman layer. We will start with the well-known relations for the surface (Prandtl) layer.

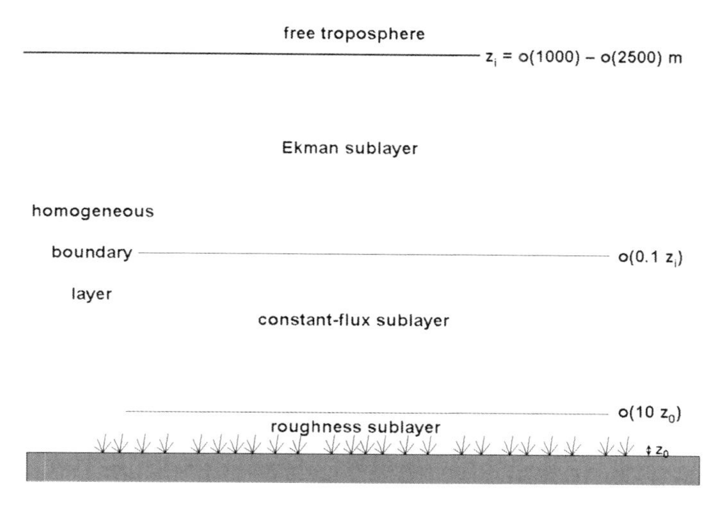

Fig. 3.1 Vertical layering in the atmospheric boundary layer over flat homogeneous terrain

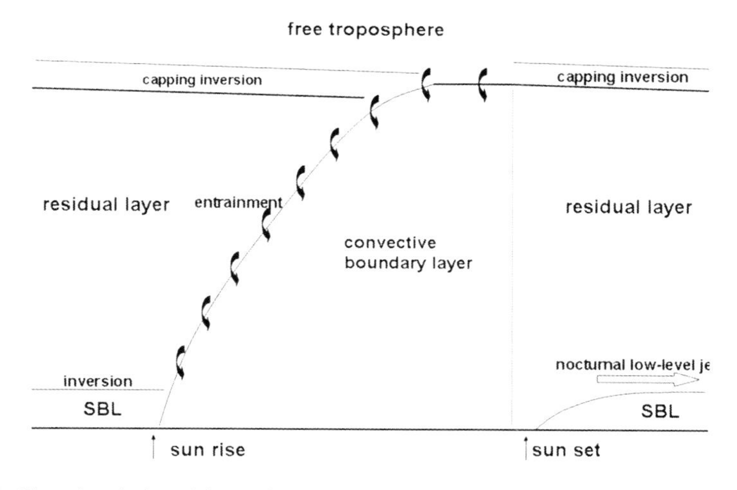

Fig. 3.2 Diurnal variation of the vertical structure of the ABL over flat terrain from noon to noon

3.1 Surface Layer (Prandtl Layer)

Smaller and older onshore wind turbines with tip heights below about 100 m are usually fully immersed in the Prandtl layer. The siting of these turbines requires knowledge mainly from the quite simple and well-known relations given below in this Sect. 3.1. The siting of larger turbines with tip heights above about 100 m requires information on the wind laws in the Ekman layer as well. These features are presented in the subsequent Sect. 3.2.

The Prandtl layer or surface layer or constant-flux layer is defined meteorologically as that layer where the turbulent vertical fluxes of momentum, heat, and moisture deviate less than 10 % from their surface values, and where the influence of the Coriolis force is negligible. Usually this layer covers only 10 % of the whole ABL depth. Although this definition seems to be a paradox because the turbulent vertical fluxes have their largest vertical gradients just at the surface, the concept of the constant-flux layer has proven to be a powerful tool to describe the properties of this layer.

We start to derive the basic wind equations for this layer by stipulating a vertically constant momentum flux, i.e. assuming a stationary mean flow in x-direction and horizontal homogeneity [no derivatives neither in wind (x) nor in cross-wind (y) direction]. This simplifies the equations of motion (2.2)–(2.4) to:

$$K_M \frac{\partial u}{\partial z} = const = u_*^2 \tag{3.1}$$

where u_* is the friction velocity defined in (3.2) and K_M is the vertical turbulent exchange coefficient for momentum, which has the effect and the physical dimension of a viscosity. K_M appears when replacing F_x in term VII of Eq. (2.2) using $F_x = \partial/\partial z\,(K_M\,\partial u/\partial z)$. A specification of K_M for neutral stratification is given at the beginning of Sect. 3.1.1.1 and for non-neutral stratification in Eq. (6.9). The friction velocity can be estimated from measured logarithmic wind profiles by inversion of Eq. 3.4 or can be derived from high-resolution wind fluctuation measurements with a sonic anemometer in the Prandtl layer:

$$u_* = \left(\overline{u'w'}^2 + \overline{v'w'}^2\right)^{\frac{1}{4}} \tag{3.2}$$

where u′ denotes the 10 Hz turbulent fluctuation of the West–East wind component, v' the fluctuation of the South-North component, and w′ the fluctuation of the vertical component.

In cases where high-resolution turbulent fluctuation measurements and wind profile data are unavailable, the friction velocity can also be inferred from the geostrophic drag law which relates the friction velocity u_* with the modulus, G of the geostrophic wind speed [see Eqs. (2.5) and (2.6)] that represents the large-scale pressure gradient force. The geostrophic drag law reads (Zilitinkevich 1975):

$$C_D = \frac{u_*}{G} = \frac{\kappa}{\sqrt{\left(\ln\frac{u_*}{fz_0} - A\right)^2 + B^2}} = \frac{\kappa}{\sqrt{\left(\ln\frac{G}{fz_0} + \ln C_D - A\right)^2 + B^2}} \tag{3.3}$$

where C_D is the geostrophic drag coefficient, z_0 is the roughness length of the surface introduced in Eq. (3.6) and A and B are two empirical parameters which principally depend on the thermal stability of the atmosphere (see Zilitinkevich 1975; Hess and Garratt 2002 or Peña et al. 2010b for details). The friction velocity computed from (3.3) is a large-scale averaged friction velocity, because the geostrophic wind speed is a large-scale feature representing a horizontal scale of the

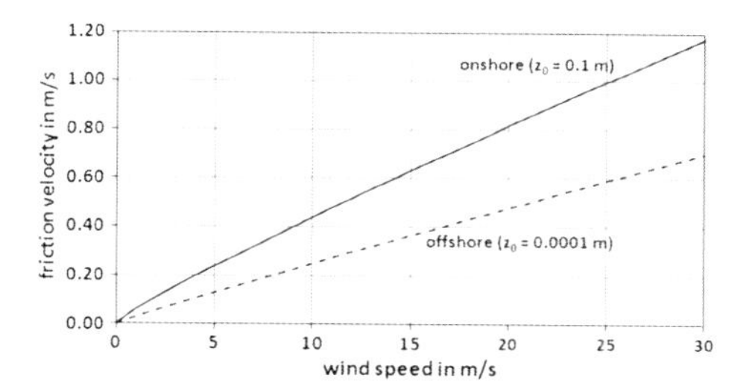

Fig. 3.3 Relation between the geostrophic wind speed, G and the friction velocity, u^* using the simplified geostrophic drag law (3.4) with $A^* = 3.8$ for onshore and $A^* = 4.7$ for offshore conditions

order of about 100 km. Unfortunately, Eq. 3.3 is an implicit relation, because the friction velocity appears on both sides of the equal sign. Therefore, simplifications of this drag law have been suggested, e.g., by Jensen (1978). Here we suggest a similar simplification which has also been used in Emeis and Frandsen (1993). Neglecting B and forming a new parameter $A^* = A - \ln C_D$ gives:

$$\frac{u_*}{G} = \frac{\kappa}{\ln \frac{G}{f z_0} - A^*} \tag{3.4}$$

Equation 3.4 can easily be solved for the friction velocity if the modulus of the geostrophic wind speed, G and the parameter A* are known. Due to the given choice of A*, the parameter A* depends on stability and on surface roughness.

A*, A and B are empirical parameters which have to be estimated from measurement data. Hess and Garratt (2002) have listed several estimations. They suggest, as the best approximation to steady, homogeneous, neutral, barotropic (no thermal wind) atmospheric conditions that they could find, i.e., the near-neutral, near-barotropic ABL in middle and high latitudes, to choose $A = 1.3$ and $B = 4.4$. Using these two values, we get $A^* = 3.7$ for a roughness length of 0.1 m (onshore) and $A^* = 4.5$ for a roughness length of 0.0001 m (offshore). Peña et al. (2010a, b) choose $A = 1.7$ and $B = 5$ to be close to the values used by the wind atlas program WAsP (Troen and Petersen 1989). This gives $A^* = 3.8$ for onshore and $A^* = 4.7$ for offshore conditions. The difference between onshore and offshore conditions using the simplified drag law (3.4) is illustrated in Fig. 3.3.

Please note that the parameters G and f are external parameters in the drag law (3.3) and its simplification (3.4). This means, that neither the drag law (3.3) or its simplification (3.4) can be used to compute a roughness length-dependent modulus of the geostrophic wind speed. As already stated in Sect. 2.3, the geostrophic wind solely depends on the large-scale horizontal pressure gradient and the latitude-dependent Coriolis parameter, but not on surface properties.

The friction velocity obtained from (3.2) or (3.3) is the usual scaling velocity for the wind speeds and the vertical wind shear in the atmospheric surface layer. In cases with strong convective vertical motions, the convective velocity scale [see (3.20)] should be used as scaling velocity.

3.1.1 Logarithmic Wind Profile

The most important atmospheric feature which influences the generation of energy from the wind is the vertical increase of wind speed with height. This increase is described by the laws for the vertical wind profile. Different descriptions of the vertical wind profile exist. We will have a look at the classical logarithmic wind profile first, which can be derived from simple physical considerations valid for the surface layer. The empirical power law, which is often used instead of the logarithmic law, will be presented in the subsequent Sect. 3.1.2.

3.1.1.1 Neutral Stratification

We start the derivation of the logarithmic wind profile with dynamical considerations, which suggest formulating the vertical momentum exchange coefficient K_M in (3.1) as being proportional to the mixing length $l = \kappa z$, which in turn is proportional to the distance to the ground and the friction velocity ($K_M = \kappa u_* z$). This leads to the following equation for the vertical wind speed gradient (or wind shear) in the Prandtl layer derived from Eq. (3.1) (with the van Kármán constant $\kappa = 0.4$):

$$\frac{\partial u}{\partial z} = \frac{u_*}{l} = \frac{u_*}{\kappa z} \tag{3.5}$$

Integration of the wind shear Eq. (3.5) from a lower height z_0 where the wind speed is assumed to vanish near the ground up to a height z within the Prandtl layer then yields the well-known logarithmic wind profile for this layer with the roughness length z_0:

$$u(z) = \frac{u_*}{\kappa} \ln \frac{z-d}{z_0} \tag{3.6}$$

where d is called the displacement height and is relevant for flows over forests and cities (see Sects. 3.6 and 3.7). The displacement height gives the vertical displacement of the entire flow regime over areas which are densely covered with obstacles such as trees or buildings. Otherwise, we will disregard this parameter in the following considerations. If the displacement height is a relevant parameter, then in the following equations all dimensionless ratios z/z_0 and z/L_* [see (3.10) for the definition of L_*] have to be replaced by $(z-d)/z_0$ and $(z-d)/L_*$ respectively.

The roughness length, z_0 and the displacement height, d are not purely local values. They depend in a non-linear way from the surface properties upstream of the place where the wind profile has to be computed from (3.6). The size of this influencing upstream area, which is called fetch or footprint, is increasing with increasing height z in the wind profile. Thus, the determination of these two values is not an easy task, but requires the operation of footprint models (Schmid 1994; Foken 2012). The footprint increases with wind speed, decreases with increasing turbulence, and increases with the measurement height. If no detailed information is available, a rough first guess of the upstream extent of the footprint is one hundred times the height z. The extent is modified by the thermal stability of the surface layer. For unstable stratification the footprint is closer to the site of interest, while for stable stratification it is further away. This means that surface features such as hills and forests can influence the wind speed and profile at hub heights in the order of 100 m even if they are several kilometres upstream. To guarantee a good representativity of an estimated wind profile from (3.6) for a certain surface type, the footprint should be horizontally as homogeneous as possible. The transfer of the footprint concept to inhomogeneous terrain is discussed in Schmid (2002).

The wind speed increases with height without a turning of the wind direction in the Prandtl layer. A scale analysis gives for the height of this layer, z_p (Kraus 2008):

$$z_p \approx 0.01 \frac{u_*}{f} \approx 0.00064 \frac{|v_g|}{f} \tag{3.7}$$

Putting in numbers ($u_* = 0.5$ m/s, $f = 0.0001$ 1/s, $v_g = 8$ m/s) gives a typical height of the Prandtl layer of 50 m.

In a well-mixed Prandtl layer the temperature, T decreases with height according to the adiabatic lapse rate, g/c_p (g is gravity acceleration, c_p is specific heat of the air at constant pressure). This yields a vertical temperature decrease of roughly 1 K per 100 m in an unsaturated atmosphere, i.e., in an atmosphere in which no moisture condensation or evaporation processes take place. Due to this vertical decrease, the normal temperature is not appropriate to identify air masses. For air mass identification, meteorologists and physicists have developed the definition of an artificial temperature which stays constant during vertical displacements without condensation processes. This artificial temperature is the potential temperature. The potential temperature,

$$\Theta = T\left(\frac{p_0}{p}\right)^{\frac{R}{c_p}} \tag{3.8}$$

is constant with height in a neutrally stratified Prandtl layer (R is the gas constant for dry air). p_0 is the surface pressure.

Equations (3.5), (3.6) and (3.8) describe vertical profiles of mean variables in the surface layer. We also have to specify the vertical distribution of turbulence. The standard deviations [see Eq. (A.4) in Appendix A] of the 10 Hz turbulent fluctuations of the three velocity components are assumed to be independent of height in the surface layer and scale with the friction velocity u_* as well (Stull 1988; Arya 1995). Usually the following relations are used:

$$\frac{\sigma_u}{u_*} \approx 2.5; \quad \frac{\sigma_v}{u_*} \approx 1.9; \quad \frac{\sigma_w}{u_*} \approx 1.3; \tag{3.9}$$

Relating the standard deviations to the mean wind speed rather than to the friction velocity leads to the definition of turbulence intensity. The streamwise turbulence intensity is defined in equation (A.6) in Appendix A. By inserting the leftmost relation from (3.9) into (3.6) we get for this turbulence intensity (Wieringa 1973):

$$I_u(z) = \frac{\sigma_u}{u(z)} = \frac{1}{\ln(z/z_0)} \tag{3.10}$$

This means that turbulence intensity in the neutrally-stratified surface layer is a function of surface roughness only. Increasing roughness lengths will lead to higher turbulence intensities. For a given roughness length, turbulence intensity decreases with height in the surface layer.

3.1.1.2 Unstable Stratification

The thermal stratification of the surface layer is rarely found to be absolutely neutral. In most cases, there is a non-vanishing virtual potential heat flux $\overline{\Theta_v' w'}$at the ground [see the definition of potential temperature in Eq. (3.8), the virtual potential temperature, Θ_v includes the modifying influence of the atmospheric humidity on the static stability of the air, see (2.14)], which leads to a thermal stratification of the surface layer. From this surface heat flux and the friction velocity u_* a length scale, L_*, the Obukhov length (sometimes also called Monin–Obukhov length, but the first term is historically more correct and will be used here) can be formed:

$$L_* = \frac{\Theta_v}{\kappa g} \frac{u_*^3}{\overline{\Theta_v' w'}} \tag{3.11}$$

The heat flux is counted positive if it is directed from the atmosphere towards the ground (cooling the atmosphere) and negative if it is towards the atmosphere (heating the atmosphere). Thus an unstable surface layer is characterized by a negative Obukhov length. The virtual potential heat flux $\overline{\Theta_v' w'}$ can be separated into a sensible heat flux and a humidity flux:

$$\overline{\Theta_v' w'} = \overline{\Theta' w'} + 0.61 \Theta \overline{q' w'} \tag{3.12}$$

The ratio of the turbulent sensible heat flux and humidity flux is called Bowen ratio, B:

$$B = \frac{c_p \overline{\Theta' w'}}{L_v \overline{q' w'}} \tag{3.13}$$

where q is specific humidity and L_v is the (latent) heat of vaporisation. The buoyancy exerted by the vertical heat and humidity gradients is given by

$\frac{g}{\Theta}\overline{\Theta' w'} + 0.61 g\overline{q' w'}$. The ratio of these two contributing terms is called the buoyancy ratio, *BR* which is inversely proportional to the Bowen ratio *B*:

$$BR = \frac{0.61\Theta\overline{q'w'}}{\overline{\Theta'w'}} = \frac{0.61 c_p \Theta}{L_v}\frac{1}{B}. \tag{3.14}$$

This unstable type of the surface layer is usually found during daytime over surfaces heated by insolation and over waters which are warmer than the air above. The degree of instability is described by the non-dimensional ratio of the height z over the Obukhov length L_*. In an unstable surface layer, warm air bubbles rise from the surface to the top of the unstable layer, which is usually marked by a temperature inversion. The height of the unstable surface layer is designated by z_i. The temperature decreases with height according to the adiabatic lapse rate, but in a shallow super-adiabatic layer near the surface the lapse rate is even stronger.

While z is the only scaling length scale in a neutrally stratifies surface layer, the modulus of the Obukhov length L_* [see (3.11)] is an additional length scale in the unstable surface layer. A non-dimensional parameter can be formed from these two length scales. The ratio z/L_* is used in the following as a stability parameter. This parameter is negative for unstable stratification, positive for stable stratification and zero for neutral stratification.

For small negative values of z/L_* the vertical wind profiles in the surface layer can be described by introducing a correction function $\Psi_m(z/L_*)$ (Paulson 1970; Högström 1988):

$$\Psi_m = 2\ln\left(\frac{1+x}{2}\right) + \ln\left(\frac{1+x^2}{2}\right) - 2arctg(x) + \frac{\pi}{2} \tag{3.15}$$

where $x = (1-b\,z/L_*)^{1/4}$ and $b = 16$. This leads to the following description of the vertical wind profile which replaces Eq. (3.6):

$$u(z) = u_*/\kappa(\ln(z/z_0) - \Psi_m(z/L_*)) \tag{3.16}$$

While the surface layer mean wind profile in the unstable surface layer depends on the local stability parameter z/L_*, turbulence partly depends on non-local parameters as well. The non-local parameter boundary layer height z_i (see also Appendix B) is another length scale in the unstable surface layer, if the thermally induced vertical motions extend through the whole depth of the convective boundary layer. This allows for the formulation of a second non-dimensional stability parameter, z_i/L_*. The standard deviations of the 10 Hz fluctuations of the wind components in the unstably stratified Prandtl layer depend either on this second parameter z_i/L_* or the first parameter z/L_* (Panofsky et al. 1977; Arya 1995):

$$\frac{\sigma_u}{u_*} = \left(15.625 - 0.5\frac{z_i}{L_*}\right)^{1/3};\ \frac{\sigma_v}{u_*} = \left(6.859 - 0.5\frac{z_i}{L_*}\right)^{1/3} \tag{3.17}$$

$$\frac{\sigma_w}{u_*} = 1.3\left(1 - 3\frac{z}{L_*}\right)^{1/3} \tag{3.18}$$

This means that the standard deviations of the horizontal wind components Eq. (3.17) are height independent in the unstable surface layer while the standard deviation of the vertical wind component (3.18) increases with height. Originally, Panofsky et al. (1977) and Arya (1995) have given 12 as a common value for the numbers 15.625 and 6.859 in (3.17). The different choice has been made here in order to be consistent with the relations in (3.9) in the limit of neutral stratification.

Arya (1995) gives for the standard deviations of the 10 Hz fluctuations of the wind components in the unstably stratified Ekman layer above the Prandtl layer:

$$\sigma_{u,v,w} = 0.6w_* \tag{3.19}$$

with the convective velocity scale:

$$w_* = \frac{gz_i}{\Theta}\overline{w'\Theta'} \tag{3.20}$$

This convective velocity scale substitutes the friction velocity as a scaling velocity in situations where vertical velocities due to unstable thermal stratification are in the same order as the horizontal wind speeds. This means that the standard deviation of the vertical velocity component increases with height in the unstable Prandtl layer due to relation (3.18) and then stays constant above it due to relation (3.19).

3.1.1.3 Stable Stratification

The stable type of the surface layer, which is characterized by a downward surface heat flux ($L_* > 0$) and a stable thermal stratification of the air, is usually found at night time, over waters that are colder than the air above, and over ice and snow-covered surfaces. For positive values of z/L_*, the correction functions for the logarithmic wind profile read (Businger et al. 1971; Dyer 1974; Holtslag and de Bruin 1988):

$$\Psi_m(z/L_*) = \begin{cases} -az/L_* & \textit{for}\ 0 < z/L_* \le 0.5 \\ Az/L_* + B(z/L_* - C/D) & \\ \exp(-Dz/L_*) + BC/D & \textit{for}\ 0.5 \le z/L_* \le 7 \end{cases} \tag{3.21}$$

where $a = 5$, $A = 1$, $B = 2/3$, $C = 5$, and $D = 0.35$. The vertical wind profile $u(z)$ in the stable surface layer is then described again by Eq. (3.16) but now using the functions (3.21) for stable stratification.

The air temperature in the stable boundary layer vertically decreases less than the adiabatic lapse rate and the potential temperature (3.8) increases with height. The standard deviations of the wind components are usually assumed to be constant with height in the same way as described by (3.9) for the neutral ABL (Arya 1995).

3.1.2 Power Law

Sometimes, instead of the logarithmic profile laws (3.6) or (3.16), which have been derived from physical and dimensional arguments, an empirical power law is used to describe the vertical wind profile:

$$u(z) = u(z_r)\left(\frac{z}{z_r}\right)^a \tag{3.22}$$

where z_r is a reference height and a is the power law exponent (sometimes called the "Hellmann exponent"). The exponent a depends on surface roughness and the thermal stability of the Prandtl layer. The analysis of the relationship between the logarithmic law (3.6) or (3.16) and the power law (3.22) is not easy, because thermal stability is described quite differently in both formulations. The following section shows how (3.6) or (3.16) and (3.22) are related to each other and whether they can be used really interchangeably.

3.1.3 Comparison Between Logarithmic and Power Law

The choice of a suitable way of describing the wind profile is often made by practical arguments. Although today computer resources set nearly no limits any more to the rapid integration of complex equations, the power law (3.22) is often chosen due to its mathematical simplicity. It is often claimed that both descriptions lead more or less to the same results. A comparison of the parameters of the two profile laws for neutral stratification is given in Table 3.1.

The following analysis shows theoretically how closely the logarithmic profiles (3.6) or (3.16) can be described by a power law (Emeis 2005). This is not a new issue as Sedefian (1980) has derived theoretically how the power law exponent n depends on z/z_0 and z/L_* by equating the slopes of a logarithmic profile and a power law. As long as the height range over which the two profiles should match is small the solution given by Sedefian (1980) is practical and sufficient. One will always find a power law with an exponent n that fits to a given logarithmic profile at a given height.

However, today's tasks in wind engineering (the construction of large wind turbines and the design of high buildings) often require the extrapolation of the wind profile over considerable height intervals. For these purposes the two descriptions are only equivalent if it is possible to find a power law that fits to the logarithmic profile not only in slope but also in curvature over the respective range. The following investigation will demonstrate that this is possible only for certain combinations of surface roughness and atmospheric stability in a stably stratified boundary-layer flow. We start the analysis for the sake of simplicity with neutral stratification.

Table 3.1 Typical profile law parameters for vertical wind profiles in the ABL: roughness length z_0, power law (Hellmann) exponent a [neutral thermal stratification, see (3.22)], friction velocity u_* (neutral stratification, 10 m/s geostrophic wind) and deviation angle from the geostrophic wind direction φ. The values should be regarded as estimates only

Surface type	z_0 [m]	a	u* [m/s]	φ [degree]
Water	0.001	0.11	0.2	15–25
Grass	0.01–0.05	0.16	0.3	
Shrubs	0.1–0.2	0.20	0.35	25–40
Forest	0.5	0.28	0.4	
Cities	1–2	0.40	0.45	
Megacities	5			
Mountains	1–5		0.45	35–45

From Emeis (2001)

For the investigation of the possibility whether the profile laws (3.6) and (3.22) can describe the same wind profile over a larger height range, we need the mathematical formulation of the slope and the curvature of the wind profiles expressed by (3.6) and (3.22). The slope of the logarithmic wind profile under neutral stratification is given by the first derivative of (3.6) with respect to the vertical coordinate z:

$$\frac{\partial u}{\partial z} = \frac{1}{\kappa}\frac{u_*}{z} = \ln^{-1}\left(\frac{z}{z_0}\right)\frac{u(z)}{z} \tag{3.23}$$

and the curvature of the logarithmic profile follows by taking the second derivative of (3.6) with respect to the vertical coordinate:

$$\frac{\partial^2 u}{\partial z^2} = -\frac{1}{\kappa}\frac{u_*}{z^2} = -\ln^{-1}\left(\frac{z}{z_0}\right)\frac{u(z)}{z^2} \tag{3.24}$$

The slope of the power law by differentiating (3.22) with respect to the vertical coordinate yields:

$$\frac{\partial u}{\partial z} = \frac{u(z_r)}{z_r} a\left(\frac{z}{z_r}\right)^a \left(\frac{z}{z_r}\right)^{-1} = au(z_r)\left(\frac{z}{z_r}\right)^a = a\frac{u(z)}{z} \tag{3.25}$$

and the curvature of the power laws reads after computing the second derivative of (3.22) with respect to the vertical coordinate:

$$\frac{\partial^2 u}{\partial z^2} = a(a-1)u(z_r)\left(\frac{z}{z_r}\right)^a \frac{1}{z^2} = a(a-1)\frac{u(z)}{z^2} \tag{3.26}$$

Equating the slopes of the logarithmic profile (3.23) and that of the power law (3.25) delivers a relation between the Hellmann exponent and the surface roughness length:

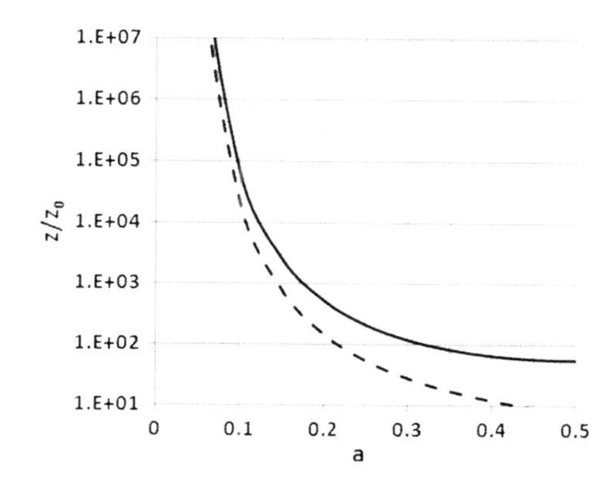

Fig. 3.4 Dependence of the power law exponent a on height and surface roughness following (3.27) (*dashed line*) and (3.28) (*full line*)

$$a = \ln^{-1}\left(\frac{z}{z_0}\right) \tag{3.27}$$

which equals the formulation given by Sedefian (1980) in the limit of neutral stratification. Comparison with the definition of the turbulence intensity (3.10) reveals that the exponent a is equal to the turbulence intensity for neutral stratification. This means, that a logarithmic wind profile and a power law profile have the same slope at a given height if the power law exponent equals the turbulence intensity at this height. Equation (3.27) is plotted in Fig. 3.4.

Equation (3.27) implies that the exponent a decreases with height for a given roughness length z_0. The height in which the slopes of the two wind profiles (3.6) and (3.22) should be equal—this is usually the anemometer height $z = z_A$—has therefore to be specified a priori. The dependence of the power law exponent a on height is stronger the smaller the ratio z/z_0 is (see Fig. 3.4). Due to this fact, the dependence of the exponent a on height is stronger for complex terrain where the roughness length z_0 is large and it can nearly be neglected for water surfaces with very small roughness lengths.

In order to see whether we can find an exponent a so that both the slope and the curvature agree in a given height we must equate the formulas (3.24) and (3.26) for the curvature of the two profiles. This yields a second relation between the Hellmann exponent and the surface roughness length:

$$a(a-1) = -\ln^{-1}\left(\frac{z}{z_0}\right) \tag{3.28}$$

For low heights over rough surfaces with $z/z_0 < 54.6$ Eq. (3.28) has no solution at all (see the full line in Figure 3.4). For $z/z_0 = 54.6$ it has one solution ($a = 0.5$) and for greater heights over smoother surfaces with $z/z_0 > 54.6$ it has two solutions of which we always choose the smaller one. This solution approaches the solution of Eq. (3.27) asymptotically as z/z_0 tends to infinity (for very smooth surfaces such as still water surfaces). Therefore, a power law with equal slope and curvature as the logarithmic profile can only exist in the limit for perfectly smooth surfaces

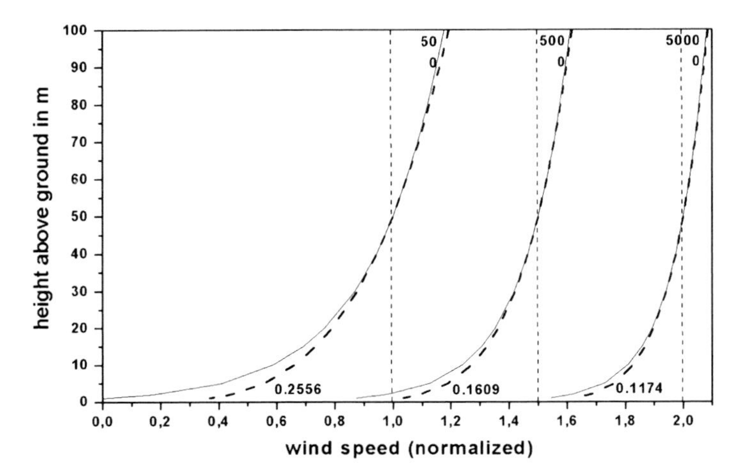

Fig. 3.5 Three normalised neutral wind profiles extrapolated from the 50 m wind speed for increasingly smooth terrain (from *left* to *right*). *Full lines*: logarithmic profiles from (3.6) (uppermost number gives z/z_0), *dashed lines*: power profiles from (3.22) (number at *bottom line* gives the exponent a). The middle curve has been shifted to the right by 0.5 and the right curve has been shifted to the right by 1.0 for better visibility

when a tends to zero. Thus, for neutral stratification, a power law with a slope and curvature that fits the logarithmic profile over a larger height range can never be constructed. Such a fit would be possible only if the power law exponent were not constant but varied with height according to (3.28).

The use of (3.27) for calculating the exponent a of a power wind profile that is an approximation to the logarithmic wind profile is better the larger z/z_0 is, i.e. the smoother the surface is. For complex terrain on the other hand, the power law with an exponent a given by (3.27) is not a good approximation to the true wind profile. This is demonstrated in Fig. 3.5 where we present wind profiles computed from (3.6) and (3.22) for three different height-to-roughness ratios z/z_0. The height where the profiles should be identical is chosen to be 50 m and the wind profiles have been normalized to the wind speed in this height. The wind speed difference between the logarithmic profile and the power law profile at 100 m height is 1.3 % for $z/z_0 = 50$ [power law exponent $a = 0.2556$ from (3.27)] and 0.3 % for $z/z_0 = 5{,}000$ ($a = 0.1174$). The relative difference between the two profiles at 10 m height is 11.2 and 2.0 % respectively.

Usually—except for very strong winds—the atmosphere is not stratified neutrally. For non-neutral stratification the slope of the logarithmic profile (3.16) is determined by:

$$\frac{\partial u}{\partial z} = \left(\ln\left(\frac{z}{z_0}\right) - \Psi\left(\frac{z}{L_*}\right)\right)^{-1} \frac{u(z)}{z}\frac{1}{x} \quad \text{for} \frac{z}{L_*} < 0 \tag{3.29}$$

$$\frac{\partial u}{\partial z} = \left(\ln\left(\frac{z}{z_0}\right) + 4.7\left(\frac{z}{L_*}\right)\right)^{-1} \frac{u(z)\left(1+4.7\frac{z}{L_*}\right)}{z} \quad \text{for} \frac{z}{L_*} > 0 \tag{3.30}$$

where the first Eq. (3.29) is valid for unstable stratification and the second Eq. (3.30) for stable stratification. The curvature of the diabatic wind profile (3.16) is given by:

$$\frac{\partial^2 u}{\partial z^2} = -\left(\ln\left(\frac{z}{z_0}\right) - \Psi\left(\frac{z}{L_*}\right)\right)^{-1} \frac{u(z)}{z^2} \frac{1 + \frac{z}{x}\frac{\partial x}{\partial z}}{x} \quad \text{for} \frac{z}{L_*} < 0 \tag{3.31}$$

$$\frac{\partial^2 u}{\partial z^2} = -\left(\ln\left(\frac{z}{z_0}\right) + 4.7\left(\frac{z}{L_*}\right)\right)^{-1} \frac{u(z)}{z^2} \quad \text{for} \frac{z}{L_*} > 0 \tag{3.32}$$

where, once again, the first Eq. (3.31) is valid for unstable stratification and the second Eq. (3.32) for stable stratification. The expression $(z/x)\ \partial x/\partial z$ in (3.31) equals $-3.75\ z/L_*\ (1/x^4)$, where x has been defined after (3.15). Slope and curvature of the power law (3.22) do not depend explicitly on stratification and thus remain unchanged. Looking for equal slopes in non-neutrally stratified flow now requires the investigation of the possible identity of (3.29)/(3.30) and (3.25). We get:

$$a = \left(\ln\left(\frac{z}{z_0}\right) - \Psi\left(\frac{z}{L_*}\right)\right)^{-1} \frac{1}{x} \quad \text{for} \frac{z}{L_*} < 0 \tag{3.33}$$

$$a = \left(\ln\left(\frac{z}{z_0}\right) + 4.7\left(\frac{z}{L_*}\right)\right)^{-1} \left(1+4.7\frac{z}{L_*}\right) \quad \text{for} \frac{z}{L_*} > 0 \tag{3.34}$$

which are exactly the equations found by Sedefian (1980) From (3.33) and (3.34) it is obvious that a is smaller with unstable stratification than with neutral, but is larger with stable stratification, because x and the expression in brackets containing z/L_* are both larger than unity.

We had seen from Fig. 3.5 that the neutral logarithmic profile is always steeper (in the manner we have plotted the Figure, steeper means that wind speed is increasing less with height) than a power law profile fitted to it at the height $z = z_A$. As the logarithmic profile for unstable stratification is even steeper than the one for neutral stratification we do not expect a match with the power law profile for unstable stratification. But for stable stratification, the slope of the logarithmic profile is smaller than for neutral conditions and a fit may become possible. We therefore equate the curvatures from Eqs. (3.26) and (3.31)/(3.32) yielding:

$$a(a-1) = -\left(\ln\left(\frac{z}{z_0}\right) - \Psi\left(\frac{z}{L_*}\right)\right)^{-1} \frac{1-3.75\frac{z}{L_*}\frac{1}{x^4}}{x} \quad \text{for} \frac{z}{L_*} < 0 \tag{3.35}$$

$$a(a-1) = -\left(\ln\left(\frac{z}{z_0}\right) + 4.7\left(\frac{z}{L_*}\right)\right)^{-1} \quad \text{for}\ \frac{z}{L_*} > 0 \tag{3.36}$$

Now, for stable stratification—in contrast to the neutral stratification above and to unstable conditions—we have the possibility to define conditions in which Eqs. (3.34) and (3.36) can be valid simultaneously. For such a power law profile which has equal slope and curvature at the height $z = z_A$, the following relation between the Hellmann exponent, a and the stability parameter, z/L_* must hold:

$$a = 1 - \left(1 + 4.7\frac{z}{L_*}\right)^{-1} \tag{3.37}$$

In contrast to the neutral case it is possible to find an exponent a for stable conditions, but this exponent depends on the static stability (expressed by z/L_*) of the flow. The possible values for a in the phase space spanned by z/z_0 and z/L_* can be found by either equating (3.34) and (3.37) or by equating (3.36) and (3.37):

$$\ln\left(\frac{z}{z_0}\right) = 2 + \frac{1}{4.7\frac{z}{L_*}} \tag{3.38}$$

Figure 3.6 illustrates the solutions from Eqs. (3.33) to (3.36), and (3.38). An evaluation of (3.38) demonstrates that the stability of the atmosphere must increase with increasing roughness and decreasing anemometer height in order to find a power law profile with the same slope and curvature as the logarithmic profile. The curved thin lines from the lower left to the upper right represent the solution of Eqs. (3.33) and (3.34), the lines with the maximum just left of $z/L_* = 0$ the solution of Eqs. (3.35) and (3.36) (please note that the lowest line is the one for $a = 0.5$, and that the lines for $a = 0.3$ and $a = 0.7$ are identical), and the thick line marks the solution of (3.38). As designed, the thick curve goes through the points where solutions from (3.34) and (3.36) are identical.

Figure 3.7 displays three examples of wind profiles for non-neutral stratification, one for unstable conditions and a large roughness length, one which lies exactly on the curve from Eq. (3.38) so that slope and curvature coincide simultaneously, and one for very stable conditions. For a roughness length of $z_0 = 0.023$ m ($z/z_0 = 2{,}173$) and a Obukhov length of $L_* = 1{,}500$ m ($z/L_* = 0.0333$) a power law profile with $a = 0.15$ has equal slope and curvature at $z = z_A = 50$ m as the logarithmic profile. At $z = 100$ m the two profiles only differ by 0.1 %, at 10 m by 0.9 %. This is an even better fit than the fit for the neutral wind profile with $z/z_0 = 5{,}000$ in Fig. 3.4. For the two profiles under unstable conditions the respective deviations at 100 m and at 10 m are 4.5 and 89.9 %, for the two profiles under very stable conditions these deviations are −3.5 and −14.0 %.

This extension of Sedefian's (1980) analysis has shown that only for certain conditions in stably stratified boundary-layer flow is it possible to find a power law profile that has the same slope and curvature as a logarithmic wind profile and thus fits the logarithmic profile almost perfectly over a wide height range. In a purely

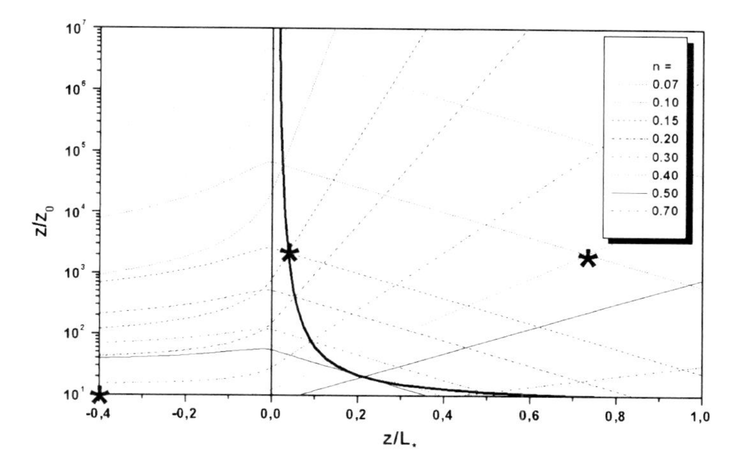

Fig. 3.6 Solution of the Eqs. (3.33)–(3.36) and (3.38) in the phase space spanned by the roughness parameter z/z_0 and the stratification parameter $z/L*$. Thin lines from lower *left* to upper *right* (calculated from (3.33) and (3.34)) indicate for different exponents a (given in the box to the upper *right*) when a logarithmic profile and a power law profile have equal slopes, thin lines from left to lower *right* (calculated from (3.35) and (3.36)) indicate for different exponents a when a logarithmic profile and a power law profile have equal curvatures, the thick line (calculated from (3.38)) runs through the points where the solutions from (3.34) and (3.36) are equal. The three asterisks mark the position of the examples shown in Fig. 3.7

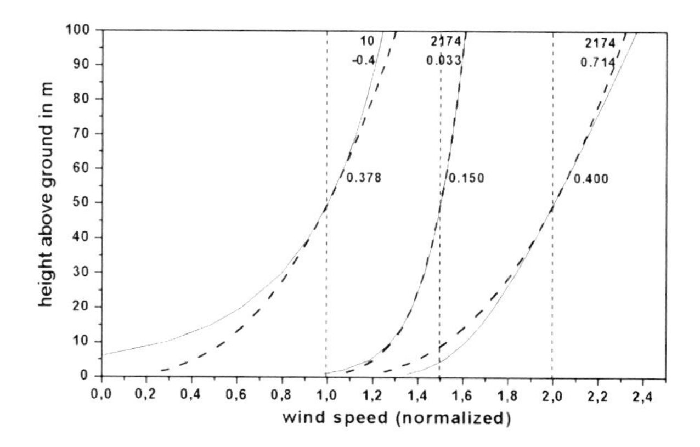

Fig. 3.7 Three normalised non-neutral wind profiles extrapolated from the 50 m wind speed for increasing stability (from *left* to *right*, the second number from above gives $z/L*$). *Full lines* logarithmic profiles from (3.16) (uppermost number gives z/z_0), *dashed lines* power profiles from (3.22) (number in the middle gives the exponent a). The middle curve has been shifted to the right by 0.5 and the right curve has been shifted to the right by 1.0 for better visibility

neutrally stratified boundary layer, this perfect fit is not possible although the fit becomes the better the smoother the surface is. The worst fit occurs for unstable conditions and rough terrain. Due to the fact that the atmosphere is usually stably stratified in the mean, it becomes obvious from the above calculations why the power law approach has been so successful in many cases.

For high wind speeds which are most favourable for wind energy conversion the stratification of the boundary-layer usually becomes nearly neutral. The above considerations then show that only for very smooth terrain (offshore and near the coasts) the power law is a good approximation to the real surface layer wind profile. Extrapolations of the wind profile above the height of the surface layer (80–100 m) by either law (3.6) or (3.22) should be made with very great care because both laws are valid for the surface layer only (Emeis 2001).

3.1.4 Vertical Wind Profile with Large Wind Speeds

Wind profiles for strong winds are nearly always close to the neutral wind profiles, because the Obukhov length defined in (3.11) takes large absolute values and the correction terms in the profile law (3.16) remain small. Vertical gust profiles look different. Wieringa (1973) derives profile exponents for gusts which are about 45 % lower than those for the mean wind. This implies that the gust factor $G(z)$ (i.e., the ratio of the gust wind speed to the mean wind speed, see equation A.33 in Appendix A) must decrease with height which has been confirmed by Davis and Newstein (1968). The decrease can be explained by the decrease of the vertical wind speed shear with height which leads to a decreasing mechanical production of turbulence. Wieringa (1973) gives an empirical relation for the height dependence of the gust factor by stipulating $D/t = 86.6$ in Eq. (A.36) in Appendix A:

$$G(z) = 1 + \frac{1.42 + 0.3013 \ln(990/(vt) - 4)}{\ln(z/z_0)} \tag{3.39}$$

The numerical value 990 m (86.6 times 11.5 m/s / 1 s) represents the turbulent length scale underneath which the majority of the turbulence elements are found. This results in $G = 1.37$ for a roughness length of 0.03 m.

3.2 Profile Laws Above the Surface Layer

Modern large wind turbines with upper tip heights of more than about 100 m frequently operate at least partly in the Ekman layer. Therefore, wind resource and load assessment cannot be done solely with the vertical profile relations and laws given in Sect. 3.1. The more complicated wind regime in the Ekman layer is to be considered as well.

The equilibrium of forces changes when moving upward from the surface layer or Prandtl layer into the Ekman layer. In addition to the pressure gradient force and the surface friction, the Coriolis force due to the Earth's rotation becomes important here as well. This means that in a stationary Ekman layer the three terms III, V, and VII in

(2.2)–(2.4) must balance each other. This layer is named from the Swedish physicist and oceanographer W. Ekman (1874–1954), who for the first time derived mathematically the influence of the Earth's rotation on marine and atmospheric flows. A prominent wind feature which distinguishes the Ekman layer from the surface or Prandtl layer below is the turning of wind direction with height. The Ekman layer covers the major part of the ABL above the Prandtl layer (see Fig. 3.1). In the Ekman layer the simplifying assumption is made that the height dependent growth of the exchange coefficient $K_M = \kappa u_* z$ (see Sect. 3.1.1.1) stops at the top of the Prandtl layer and that K_M is vertically constant for the rest of the boundary layer.

3.2.1 Ekman Layer Equations

The balance of forces in the Ekman layer involves three forces. The Coriolis force is relevant in addition to the pressure gradient force and the frictional forces. Equating the three relevant terms III, V, and VII in (2.2) and (2.3) leads to:

$$-fv + \frac{1}{\rho}\frac{\partial p}{\partial x} - \frac{\partial\left(K_M \frac{\partial u}{\partial z}\right)}{\partial z} = 0 \tag{3.40}$$

$$fu + \frac{1}{\rho}\frac{\partial p}{\partial y} - \frac{\partial\left(K_M \frac{\partial v}{\partial z}\right)}{\partial z} = 0 \tag{3.41}$$

Here, the right-most terms on the left-hand side of (3.40) and (3.41) are substituted for the symbolic expressions for the frictional forces F_x and F_y in term VII in (2.2) and (2.3). K_M is the turbulent vertical exchange coefficient for momentum, which has the physical dimension of a viscosity, i.e. m^2/s. Using the definition of geostrophic winds introduced in Eqs. (2.5) and (2.6) leads to:

$$-fv + fv_g - \frac{\partial\left(K_M \frac{\partial u}{\partial z}\right)}{\partial z} = 0 \tag{3.42}$$

$$fu - fu_g - \frac{\partial\left(K_M \frac{\partial v}{\partial z}\right)}{\partial z} = 0 \tag{3.43}$$

The two left terms in (3.42) and (3.43) can be merged into one term containing the so-called velocity deficits $u_g - u$ and $v_g - v$. This yields the so-called defect laws for the Ekman layer:

$$f(v_g - v) - \frac{\partial\left(K_M \frac{\partial u}{\partial z}\right)}{\partial z} = 0 \tag{3.44}$$

$$-f\left(u_g - u\right) - \frac{\partial\left(K_M \frac{\partial v}{\partial z}\right)}{\partial z} = 0 \tag{3.45}$$

An analytical solution of (3.44) and (3.45) is possible under certain assumptions and will be described in the following Sect. 3.2.2. The derivation of the vertical wind profile for the Ekman layer will be continued in Sect. 3.2.3.

3.2.2 Inertial Oscillations in the Ekman Layer

Up to now, we have considered stationary situations which form under different equilibria of forces. An interesting instationary situation, which is quite realistic as we will show below in Sect. 3.4, is the sudden disappearance of frictional forces in the Ekman layer in the evening hours or when the winds blow from the rough land out over the very smooth sea. In such cases, the first terms in (3.44) and (3.45) are suddenly balanced only by the inertial terms I in (2.2) and (2.3), and we yield the following equations for the temporal variation of the horizontal wind components:

$$\frac{\partial u}{\partial t} = -f\left(v - v_g\right) \tag{3.46}$$

$$\frac{\partial v}{\partial t} = f\left(u - u_g\right) \tag{3.47}$$

The terms on the left-hand side involve a dependence on time. Therefore, the analytical solution of (3.46) and (3.47) describes an oscillation with time, t:

$$u - u_g = D_v \sin ft + D_u \cos ft \tag{3.48}$$

$$v - v_g = D_v \cos ft - D_u \sin ft \tag{3.49}$$

where D_u and D_v are the ageostrophic wind components at the beginning of the oscillation in the moment when the friction vanishes. This interesting phenomenon is the basis for the development of low-level jets and is discussed in more detail in Sect. 3.4.

3.2.3 Vertical Wind Profiles in the Ekman Layer

We now derive the laws for the vertical wind profile in the Ekman layer. One can solve the defect laws (3.44) and (3.45) analytically in order to obtain the vertical wind profile in the Ekman layer (Stull 1988), if we assume a vertically constant exchange coefficient K_M:

$$u^2(z) = u_g^2\left(1 - 2e^{-\gamma z}\cos(\gamma z) + e^{-2\gamma z}\right) \tag{3.50}$$

where we have introduced another, this time inverse, length scale, γ which depends on the Coriolis parameter and the turbulent viscosity, K_M:

$$\gamma = \sqrt{\frac{f}{2K_M}} \tag{3.51}$$

Usually, the top height of the Ekman layer z_g is estimated from this inverse length scale by:

$$z_g = \frac{\pi}{\gamma} \tag{3.52}$$

Equation (3.50) can be mathematically simplified if the height z is small compared to the length scale $1/\gamma$. Then the cosine-function in (3.50) is close to unity and we get:

$$u^2(z) = u_g^2\left(1 - 2e^{-\gamma z} + e^{-2\gamma z}\right) \tag{3.53}$$

and after taking the square root we end up with:

$$u(z) = u_g(1 - e^{-\gamma z}) \tag{3.54}$$

The simplified Eq. (3.54) describes an exponential approach with height of the wind speed $u(z)$ from lower wind speed values within the Ekman layer to the geostrophic wind speed u_g above the Ekman layer, while the full Eq. (3.50) describes this approach as well, but including a small oscillation of the wind speed around the geostrophic value near the top of the Ekman layer. The full Eq. (3.50) is usually preferable, because the simplified Eq. (3.54) gives wind speed values which are—compared to (3.50)—by $1/\sqrt{2}$ too low close to the ground. Equation (3.54) is introduced here, because it has been used in some examples shown in Sects. 3.4.1 and 4.2.4. Generally, neither Eq. (3.50) nor (3.54) should be extrapolated down into the surface layer. Profile relations which are valid over the surface layer and the Ekman layer are derived in the next section.

The vertical profile of the standard deviations of the wind components as the major turbulence parameter has already been given above in Eq. (3.19).

3.2.4 Unified Description of the Wind Profile for the Boundary Layer

For many purposes, especially in those situations where the hub height is close to the top of the surface layer and the rotor area of a wind turbine cuts through the surface layer and the Ekman layer above, a unified description of the wind profile for the entire lower part of the ABL is desirable, which is valid in both layers. Due to the assumption of the constant exchange coefficient K_M in the Ekman layer, the relations (3.40)–(3.45) and (3.50) cannot be extended from the Ekman layer down into the Prandtl layer. Likewise, due to the assumption of a mixing length which grows

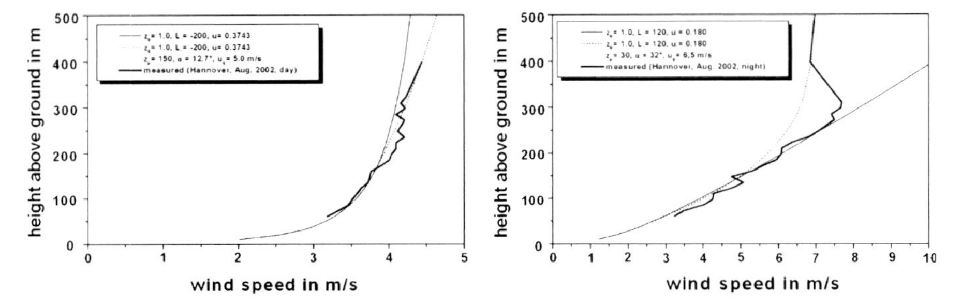

Fig. 3.8 Comparison of logarithmic wind profiles from (3.16) (*thin line*, parameters are given in the first line of the insert) and smooth boundary layer wind profiles from Eq. (3.65) (*dotted line*, parameters are given in the second and third line of the insert) with measured monthly mean data (*bold line*). *Left* daytime profiles, *right* night-time profiles, both from SODAR data for August 2002 in Hannover (Germany). (From Emeis et al. 2007b)

linearly with height in the surface layer, the logarithmic relations (3.6) and (3.16) cannot be extended into the Ekman layer. Therefore, two approaches have been tested to overcome this problem. The first idea is to fit the Prandtl and Ekman profiles together in such a way that there is a smooth transition in terms of wind speed and wind shear between both regimes. The second idea is to modify the mixing length in order to extrapolate the Prandtl layer wind profile into higher layers.

Etling (2002) had proposed the first idea by presenting a wind profile description with a linearly increasing exchange coeffiicient K_M below the Prandtl layer height, z_p and a constant K_M above this height:

$$u(z) = \begin{cases} u_*/\kappa \ln(z/z_0) & for\ z < z_p \\ u_g(-\sin\alpha_0 + \cos\alpha_0) & for\ z = z_p \\ u_g\big[1 - 2\sqrt{2}e^{-\gamma(z-z_p)} & \\ \sin\alpha_0 \cos(\gamma(z - z_p) + \pi/4 - \alpha_0) & for\ z > z_p \\ +2e^{-2\gamma(z-z_p)} \sin^2\alpha_0\big]^{1/2} & \end{cases} \tag{3.55}$$

The vertical wind profile given by Eq. (3.55) depends on five parameters: the surface roughness z_0, the geostrophic wind speed u_g, the height of the Prandtl layer z_p, the friction velocity u_*, and the angle between the surface wind and the geostrophic wind α_0. The two variables z_0 and u_g are external parameters, the other three of them are internal parameters of the boundary layer. If a fixed value is chosen for z_p then two further equations are needed to determine u_* and α_0. Equation (3.55) describes a smooth transition of wind speed from the Prandtl layer to the Ekman layer (see Fig. 3.8).

Deviating from the original approach of Etling (2002) the unified vertical wind profile should be generated from the more realistic physical requirement that both the wind speed as well as the wind shear are continuous at the height $z = z_p$ (Emeis et al. 2007b). Equating the first two equations of the wind profile Eq. (3.55) for $z = z_p$ gives an equation for the friction velocity:

$$u_* = \frac{\kappa u_g(-\sin\alpha_0 + \cos\alpha_0)}{\ln(z_p/z_0)} \tag{3.56}$$

and from equating the respective equations for the vertical wind shear at the same height $z = z_p$ we get a second equation for u_*:

$$u_* = 2|u_g|\gamma\kappa z_p \sin\alpha_0 \tag{3.57}$$

These two equations must be valid simultaneously. Equating the right hand sides of these two Eqs. (3.56) and (3.57) yields the desired relation for the turning angle, α_0:

$$\alpha_0 = arctg\frac{1}{1 + 2\gamma z_p \ln(z_p/z_0)} \tag{3.58}$$

Unfortunately, Eq. (3.58) still depends on the friction velocity u_* via the definition of γ: For the height $z = z_p$ we have from the definition of the inverse length scale, γ (3.51):

$$\gamma = \sqrt{\frac{f}{2\kappa u_* z_p}} \tag{3.59}$$

Thus, the friction velocity u_* has to be determined iteratively starting with a first guess for u_* in (3.59), subsequently computing α_0 from (3.58), and then recomputing u_* from (3.56) or (3.57).

Inversely the system of Eqs. (3.56)–(3.59) can be used to determine the height of the Prandtl layer, z_p if the friction velocity, u_* is known from other sources.

The second idea is to modify the dependence of the mixing length on height and has been proposed by Gryning et al. (2007). They reformulated the height-dependent mixing length l (which is denoted $\kappa z = L_L$ in the Prandtl layer in the following equations) in order to limit its growth with height and thus to extend the validity of the logarithmic law (3.6) to above the surface layer. They have chosen:

$$\frac{1}{l} = \frac{1}{L_L} + \frac{1}{L_M} + \frac{1}{L_U} \tag{3.60}$$

A modified mixing length is formed in (3.60) by introducing a length scale for the middle part of the boundary layer, $L_M = u_*/f\ (-2\ ln(u_*/(fz_0)) + 55^{-1}$ and a length scale for the upper part of the boundary layer, $L_U = (z_i - z)$. This results in the following wind profile alternative to (3.6) or (3.55):

$$u(z) = \frac{u_*}{\kappa}\left(\ln\frac{z}{z_0} + \frac{z}{L_M} - \frac{z}{z_i}\frac{z}{2L_M}\right) \tag{3.61}$$

Peña et al. (2010a) suggest a similar approach for the mixing length starting from Blackadar's (1962) principal approach for the mixing length, l

$$l = \frac{\kappa z}{1 + \left(\frac{\kappa z}{\eta}\right)^d} \tag{3.62}$$

which can be rewritten as

$$\frac{1}{l} = \frac{1}{\kappa z} + \frac{(\kappa z)^{d-1}}{\eta^d} \tag{3.63}$$

Incorporating this approach into the logarithmic profile law (3.6) gives:

$$u(z) = \frac{u_*}{\kappa}\left(\ln\frac{z}{z_0} + \frac{1}{d}\left(\frac{\kappa z}{\eta}\right)^d - \frac{1}{1+d}\frac{z}{z_i}\left(\frac{\kappa z}{\eta}\right)^d - \frac{z}{z_i}\right) \tag{3.64}$$

For neutral stability and $d = 1$, Peña et al. (2010a) find for the limiting value of the length scale in the upper part of the boundary layer $\eta = 39$ m; for $d = 1.25$ they give $\eta = 37$ m. The only necessary parameter in (3.64) from above the surface layer is the height of the boundary layer, z_i. A summarizing paper comparing the different approaches (3.6), (3.61) and (3.64) for neutral stratification and homogeneous terrain has been written by Peña et al. (2010a).

With the correction functions for non-neutral thermal stability (3.15) and (3.21), the unified vertical wind profile (3.55) becomes:

$$u(z) = \begin{cases} u_*/\kappa(\ln(z/z_0) - \Psi_m(z/L_*)) & \mathit{for}\ z < z_p \\ u_g(-\sin\alpha_0 + \cos\alpha_0) & \mathit{for}\ z = z_p \\ u_g\left[1 - 2\sqrt{2}e^{-\gamma(z-z_p)}\right. & \\ \sin\alpha_0\cos(\gamma(z-z_p) + \pi/4 - \alpha_0) & \mathit{for}\ z > z_p \\ \left.+2e^{-2\gamma(z-z_p)}\sin^2\alpha_0\right]^{1/2} & \end{cases} \tag{3.65}$$

In the non-neutral case the equations for the friction velocity and the wind turning angle (3.56)–(3.58) take the following forms, which now involve correction functions for the thermal stability of the atmosphere:

$$u_* = \frac{\kappa u_g(-\sin\alpha_0 + \cos\alpha_0)}{\ln(z_p/z_0) - \Psi_m(z_p/L_*)} \tag{3.66}$$

$$u_* = \frac{2|u_g|\gamma\kappa z_p\sin\alpha_0}{\varphi(z_p/L_*)} \tag{3.67}$$

$$\alpha_0 = arctg\frac{1}{1 + \frac{2\gamma z_p}{\varphi(z_p/L_*)}\left(\ln(z_p/z_0) - \Psi_m(z_p/L_*)\right)} \tag{3.68}$$

u_* and α must be determined by the same iterative procedure as described after (3.59). γ still has the form given in (3.59), b is set to 16 following Högström

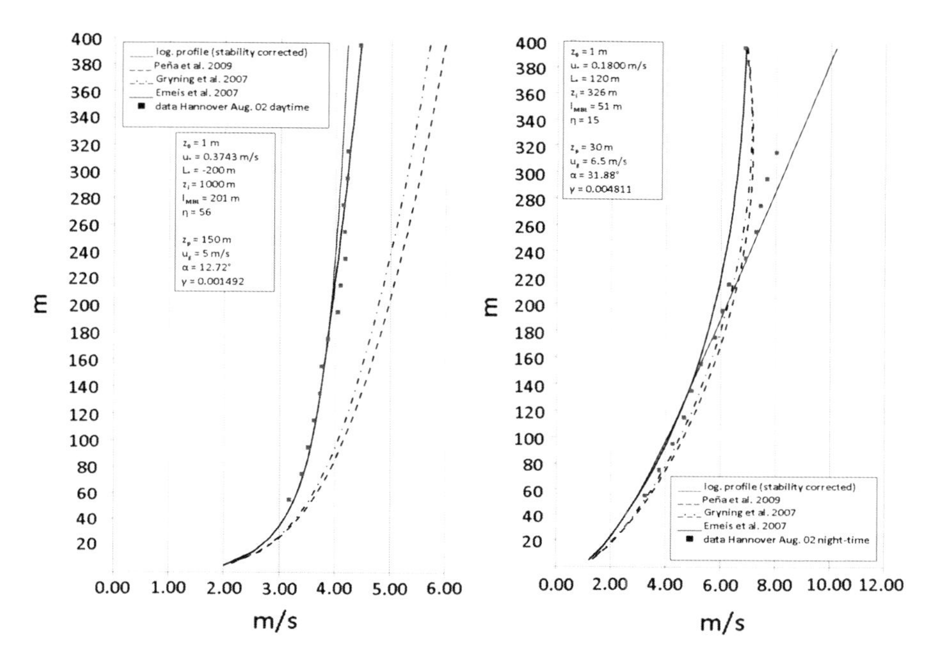

Fig. 3.9 Comparison of monthly mean vertical wind profiles from (3.16) (*thin full line*), from (3.65) (*full line*), from (3.69) (*dash-dotted line*), and from (3.70) (*dashed line*) with the same data as in Fig. 3.8. *Left* daytime, *right* night-time. The parameters for the various formulas are given in the insert

(1988). The function Ψ had been defined in (3.15) and (3.21) for unstable and stable conditions. The function φ is also different for unstable and stable stratification and is specified below in Eqs. (3.71) and (3.73).

The alternative approaches by Gryning et al. (2007) and Peña et al. (2010a, b) yield the following unified vertical wind profiles which have to be used in place of (3.61) and (3.64) in case of non-neutral thermal stratification of the boundary layer:

$$u(z) = \frac{u_*}{\kappa}\left(\ln\frac{z}{z_0} + T\left(\frac{z}{L_*}\right) + \frac{z}{L_M} - \frac{z}{z_i}\frac{z}{2L_M}\right) \tag{3.69}$$

$$u(z) = \frac{u_*}{\kappa}\left(\ln\frac{z}{z_0} + T\left(\frac{z}{L_*}\right) + \frac{1}{d}\left(\frac{\kappa z}{\eta}\right)^d - \frac{1}{1+d}\frac{z}{z_i}\left(\frac{\kappa z}{\eta}\right)^d - \frac{z}{z_i}\right) \tag{3.70}$$

with a stability correction function $T(z/L_*)$ that is again different for unstable and stable stratification. It is specified below in Eqs. (3.72) and (3.74). For details the reader is referred to Gryning et al. (2007) and Peña et al. (2010b). A comparison of wind profiles from (3.65), (3.69) and (3.70) is shown in Fig. 3.9.

For unstable situations the differential form φ of the correction function Ψ_m (3.15) for unstable thermal stratification reads:

$$\varphi(z/L_*) = (1 + \mathrm{b}\ \mathrm{z/L_*})^{-1/4} \tag{3.71}$$

and the stability correction function in (3.69) and (3.70) becomes (Peña et al. 2010b):

$$T(z/L_*) = -\Psi_m(\mathrm{z/L_*}) \tag{3.72}$$

In stable situations the differential form φ of the correction function Ψ_m (3.15) for stable stratification reads:

$$\varphi(z/L_*) = 1 + \mathrm{a}\ \mathrm{z/L_*} \tag{3.73}$$

and the stability correction function in (3.69) and (3.70) becomes (Peña et al. 2010b):

$$T(z/L_*) = -\Psi_m(\mathrm{z/L_*})\left(1 - \frac{\mathrm{z}}{2\mathrm{z_i}}\right) \tag{3.74}$$

As above, b is chosen around 15 or 16 and a is chosen to be around 4.7 and 5. The work on such unified profiles is ongoing, see. e.g., Sathe et al. (2011).

Measurement methods that are needed to determine the mixed layer height are described in Appendix B.

3.3 Spectra

Power spectra (or shortly spectra) describe the frequency dependence of the power of turbulent fluctuations, while the standard deviations of the wind components given in the relations (3.9) and (3.17) to (3.20) are integral values over the entire turbulence spectrum. The frequency dependence of the standard deviations is needed for load calculations for wind turbines. This desired information can be obtained from turbulence spectra only. Kaimal et al. (1972) give universal functions for the turbulence spectra for neutral stratification over flat terrain:

$$\frac{nS_u(n)}{u_*^2} = 105\frac{f}{(1+33f)^{5/3}} \tag{3.75}$$

$$\frac{nS_v(n)}{u_*^2} = 17\frac{f}{(1+9.5f)^{5/3}} \tag{3.76}$$

$$\frac{nS_w(n)}{u_*^2} = 2\frac{f}{(1+5.3f^{5/3})} \tag{3.77}$$

with the spectral power density $S(n)$, frequency n, and normalized frequency $f = nz/U$. The functions (3.75)–(3.77) follow Kolmogrov's –5/3 law for the inertial subrange between the low-frequency production range and the high-frequency dissipation range. Teunissen (1980) suggests a modification of these formulae for rougher terrain. He puts:

$$\frac{nS_u(n)}{u_*^2} = 105 \frac{f}{(c_u + 33f)^{5/3}} \tag{3.78}$$

$$\frac{nS_v(n)}{u_*^2} = 17 \frac{f}{(c_v + 9.5f)^{5/3}} \tag{3.79}$$

$$\frac{nS_w(n)}{u_*^2} = 2 \frac{f}{(c_w + 5.3f^{5/3})} \tag{3.80}$$

with $c_u = c_w = 0.44$ and $c_v = 0.38$ for agricultural flat terrain. The values for c_u, c_v and c_w less than unity lead to an increase of the spectral density in the low-frequency range. Alternatively, the von Kármán formulation of the spectra can be used (Teunissen 1980) which does not depend on a determination of the friction velocity but rather on the variances of the velocity components and three turbulent length scales.

$$\frac{nS_u(n)}{\sigma_u^2} = \frac{4kL_u^x}{\left(1 + 70.7(kL_u^x)^2\right)^{5/6}} \tag{3.81}$$

$$\frac{nS_v(n)}{\sigma_v^2} = 4kL_v^x \frac{1 + 188.4(2kL_v^x)^2}{\left(1 + 70.7(2kL_v^x)^2\right)^{11/6}} \tag{3.82}$$

$$\frac{nS_w(n)}{\sigma_w^2} = 4kL_w^x \frac{1 + 188.4(2kL_w^x)^2}{\left(1 + 70.7(2kL_w^x)^2\right)^{11/6}} \tag{3.83}$$

where $k = n/U$ and L_u^x, L_u^x, L_u^x are "free" scaling parameters which can be chosen to match the data. Teunissen (1980) gives

$$L_u^x = 0.146/k_u^p; \quad L_v^x = 0.106/k_v^p; \quad L_w^x = 0.106/k_w^p \tag{3.84}$$

where the k_i^p are the wave numbers of the peaks in the spectrum.

The spectra look different in non-neutral conditions. Kaimal et al. (1972) give for high frequencies $f > 4$:

$$\frac{nS_u(n)}{u_*^2 \varphi_\varepsilon^{2/3}} = \alpha_1 (2\pi\kappa)^{-2/3} f^{-2/3} \tag{3.85}$$

$$\frac{nS_v(n)}{u_*^2 \varphi_\varepsilon^{2/3}} = \frac{nS_w(n)}{u_*^2 \varphi_\varepsilon^{2/3}} = 0.4 f^{-2/3} \tag{3.86}$$

with the universal constant $\alpha_1 = 0.5$ and the non-dimensional dissipation rate for turbulent kinetic energy $\phi_\varepsilon = \kappa z\varepsilon/u_*^3$. Kaimal et al. (1972) derive from data of the Kansas experiment:

$$\varphi_\varepsilon^{2/3} = \begin{matrix} 1 + 0.5\left|\frac{z}{L_*}\right|^{2/3} & \text{for } -2 \leq \frac{z}{L_*} \leq 0 \\ 1 + 2.5\left|\frac{z}{L^*}\right|^{3/5} & \text{for } 0 \leq \frac{z}{L_*} \leq 2 \end{matrix} \tag{3.87}$$

For lower frequencies f < 4 (3.85) and (3.86) depend on z/L_* as well. According to Kaimal et al. (1972) the shape of these spectra is similar to (3.74) to (3.76).

The height dependence of the wave number k_{max} of the maximum of the spectrum that describes the lateral extension of turbulence elements has been found empirically to be (Schroers et al. 1990):

$$k_{max} = 0.0028z^{-0.27} \tag{3.88}$$

The integral length scale that describes the longitudinal extension of turbulence elements is found to vary according to (Schroers et al. 1990):

$$L_x(z) = 112.3z^{0.27} \tag{3.89}$$

This means that $L_x = 367$ m at 80 m and 389 m at 100 m height. L_x and $1/k_{max}$ are related to each other. L_x is about one third of $1/k_{max}$. Schroers et al. (1990) further found $L_x/L_y = 4.6$ at 48 m height and $L_x/L_y = L_x/L_z$ at 80 m, where L_y is the lateral and L_z the vertical extension of the turbulence elements.

3.4 Diurnal Variation of the Wind Profile

The usual daily changes in the thermal stratification of the atmospheric boundary layer over land influence vertical wind profiles as well. These wind profiles have been introduced for stationary conditions in Sects. 3.1.1.2 and 3.1.1.3. Non-stationarity provokes additional features not covered by the stationary wind laws which go beyond the necessary changes between the differently shaped wind profiles under different thermal stratification. Over oceans the diurnal cycle is practically absent due to the high heat capacity of the water. Instead, we find here an annual cycle. See Sect. 5.2 for further details.

The diurnal variation is considerably different for near-surface winds and winds above a certain height which has become known as the "reversal height" or "cross-over height". Near-surface winds under clear sky conditions behave as everyone knows from own experience: the wind freshens during daytime and calms down at night-time. The opposite is occurring above the cross-over height: wind speed is higher at night-time and decreases during daytime. This feature has already been described by Hellmann (1915) and Peppler (1921) from the evaluation of wind measurements from high masts. Wieringa (1989) has given a more general overview on this phenomenon. The term "cross-over" comes from plots displaying mean daytime and mean night-time vertical wind profiles. These two profiles cross each other at the cross-over height. Below this height, the mean daytime wind speed is larger than the mean nocturnal wind speed while above this height the opposite is true. This leads to the phenomenon that at cross-over height the diurnal variation of wind speed is at a minimum. Emeis (2004) and Emeis et al. (2007b) have demonstrated this effect from ground-based acoustic soundings with a SODAR. E.g., Emeis (2004) shows the diurnal wind variation at different heights for a rural area (Fig. 3.10). Emeis et al. (2007b) find a cross-over height of a bit more than 100 m for spring in Hannover (Germany) (Fig. 3.11). Lokoshchenko and Yavlyaeva (2008) find a cross-over height from sodar data of 60–80 m for spring and summer in Moscow.

Daytime wind speeds in both layers below and above the cross-over height are more or less equal due to the intense vertical mixing in the daytime convective boundary layer. At night-time the strong stabilisation of the boundary layer due to the radiative cooling of the ground leads to a decoupling of the winds above and below the cross-over height. Winds below this height no longer feel the driving winds from higher layers while winds above this height speed up due to the missing frictional force from below. This nocturnal speed-up above the cross-over height leads to the formation of low-level jets.

3.4.1 Vertical Profiles of the Weibull Parameters

The cross-over height, which has been introduced in the preceding section, is related to the vertical profile of the shape parameter k of Weibull distributions of the 10 min wind speeds as well (see Appendix A and Wieringa 1989), as this parameter is inversely related to the temporal variance of wind speed (see Eq. (A.28)). Thus, the vertical profile of the shape parameter must have a maximum at the cross-over height because the diurnal variation of the wind speed is at a minimum here. Evaluations in Emeis (2001) clearly show such maxima in the shape parameter profiles at heights between 60 and 80 m.

Independent from the wind profile laws which have been introduced in Sect. 3.1 above and which easily apply to the scale parameter A of the Weibull distribution as well, several empirical formulas for the vertical variation of the shape parameter k have been suggested from earlier studies. Justus et al. (1978) fitted profile functions from tower data up to 100 m a.g.l. by:

$$k(z) = k_A \frac{1 - c \ln\left(\frac{z_a}{z_{ref}}\right)}{1 - c \ln\left(\frac{z}{z_{ref}}\right)} \tag{3.90}$$

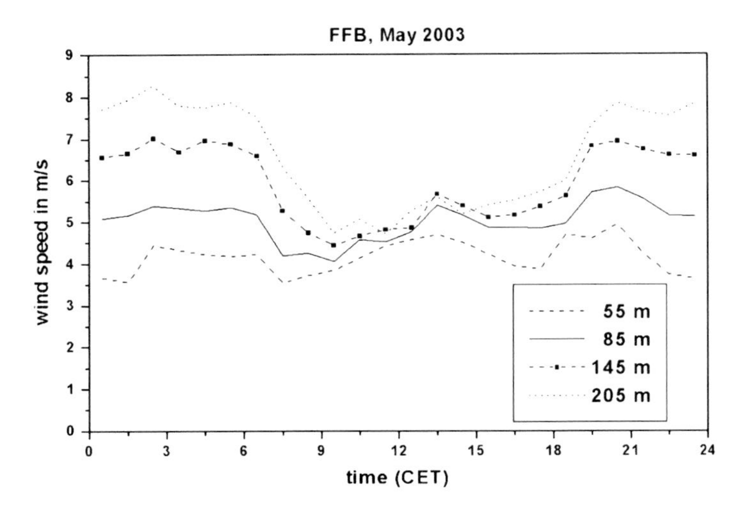

Fig. 3.10 Diurnal variation of wind speed at different heights (55, 85, 145 and 205 m) above ground from SODAR observations over a flat rural area

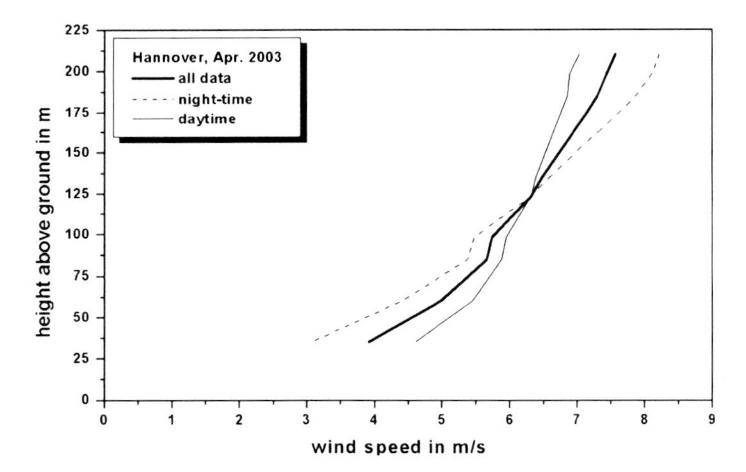

Fig. 3.11 Mean monthly vertical wind profiles from SODAR measurements for daytime (*thin full line*) and for night-time (*dashed line*) over the city of Hannover (Germany) in April 2003

with k_A as the measured shape parameter at the height z_A, $z_{ref} = 10$ m, and $c = 0.088$. Justus et al. (1978) were principally aware of the possible existence of a maximum in the k-profile but assumed that this maximum would occur at heights above 100 m. Later Allnoch (1992) proposed to put $c = 0.19$ and $z_{ref} = 18$ m in order to better represent the slope of the k-profile at the top of the surface layer.

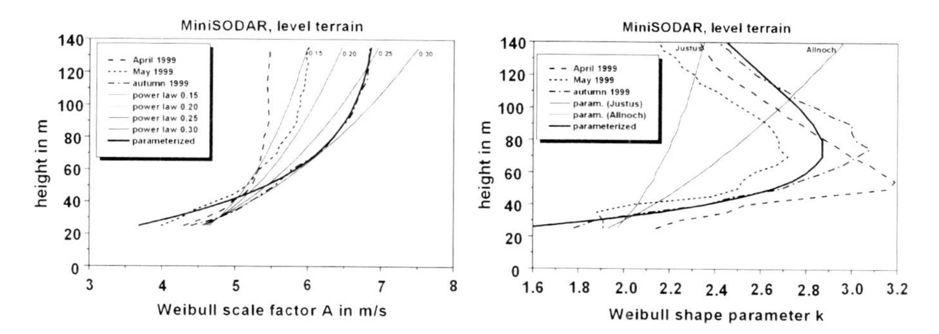

Fig. 3.12 Vertical profiles of the scale (*left*) and the shape (*right*) parameter of the Weibull distribution from SODAR measurements (differently *dashed curves*) for flat terrain. The parameterized curves on the left are from (3.22) (exponents are given at the upper end of the thin full curves), (3.54) using $A_g = 6.98$ m/s and $\gamma = 0.03$ (*bold curve*), and on the right from (3.91) using $z_m = 75$ m and $c_2 = 0.06$. The curves labelled "Justus" and "Allnoch" have been computed from (3.90)

Wieringa (1989) tried a different approach for the description of the vertical profile of the shape parameter in which he took into account the expected maximum at the top of the surface layer, although he complained that the existence of this maximum had not yet been proven until the publication of his paper. He rather parameterizes the difference $k(z) - k_A$ instead of the ratio $k(z)/k_A$ by putting:

$$k(z) - k_A = c_2(z - z_A)\exp\left(-\frac{z - z_A}{z_m - z_A}\right) \tag{3.91}$$

with the expected height of the maximum of the k-profile, z_m and a scaling factor, c_2 of the order of 0.022 for level terrain. c_2 determines the range between the maximum value of $k(z)$ at height z_m and the asymptotic value of k at large heights. Thus (3.91) contains two tunable parameters, z_m and c_2 which have to be determined from experimental data. Figure 3.12 shows examples of profile measurements with a SODAR (See Emeis 2001 for experimental details) and comparisons with the profile equations for the surface layer (3.22) and the Ekman layer (3.54). The scale parameter at heights up to about 60–80 m follows well the power law (3.22). For greater heights the Ekman law (3.50) is more suitable. Thus, Eq. (3.65) could be a good way to describe the vertical profile of the scale parameter.

We see in Fig. 3.12a distinct maximum in the vertical profile of the shape parameter at 50–80 m height. This maximum is an indication for the cross-over height. Systematic studies on the variation of the cross-over height with season and surface roughness seem to be missing. The cross-over height is different from the mixed layer height (see Appendix B). Cross-over of the daytime and night-time wind profiles usually happens together with the occurrence of low-level jets (see next section). The author's own evaluations from SODAR measurements seem to indicate that the cross-over height is roughly one third of the height of the core of the low-level jet.

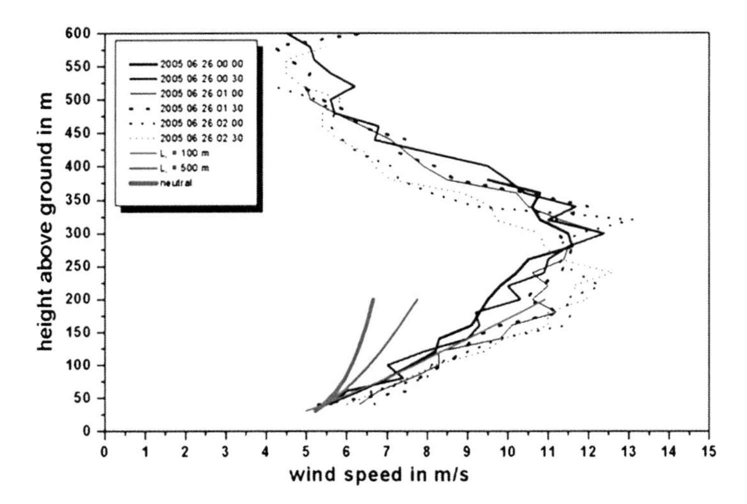

Fig. 3.13 SODAR observations of a nocturnal low-level jet over Paris Airport Charles de Gaulle in June 2005. Displayed are six consecutive half-hourly averaged wind profiles. The three curves between 30 and 200 m are from (3.16) using $L_* = \infty$, 500 and 100 m (from *left* to *right*)

3.4.2 Low Level Jets

Over land, low-level jets are nocturnal maxima in the vertical wind speed profile which form at the top of the nocturnal boundary layer. Typical heights are between 150 and 500 m above ground. Therefore, they have the ability to influence the energy yield of modern wind turbines with hub heights of more than 100 m. Figure 3.13 shows six subsequent half-hour mean profiles as an example.

3.4.2.1 Origin of Low Level Jets

The formation of low-level jets requires a temporal or spatial change in the thermal stability of the atmosphere which leads to a sudden change between two different equilibria of forces. The flow must transit from an unstable or neutral condition where friction, pressure-gradient and Coriolis forces balance each other to a stable condition where only pressure-gradient and Coriolis force balance each other (see Fig. 3.14). The sudden disappearance of the retarding friction in the equilibrium of forces leads to an inertial oscillation of the horizontal wind vector. Wind speed shoots to much higher values and the increased wind speed leads to a stronger Coriolis force which provokes a turning of the wind vector as well. The relevant equations for this phenomenon have already been presented in Sect. 3.2.2.

In the temporal domain this corresponds to a sudden change from an unstable daytime convective boundary layer to a nocturnal stable boundary layer. This requires clear skies in order to have rapid changes in thermal stratification but still

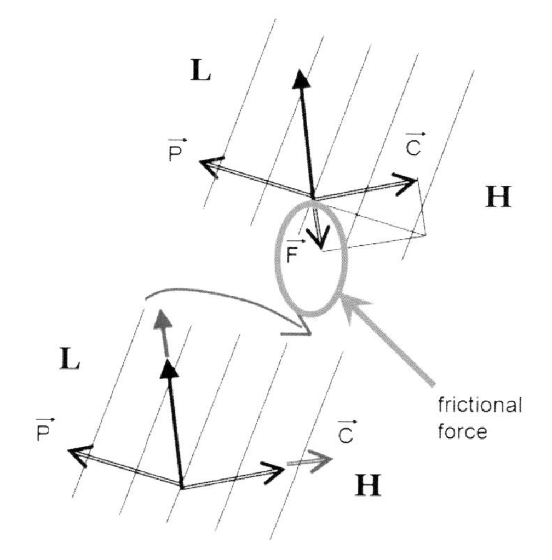

Fig. 3.14 Balance of forces (*black arrows*) in the daytime convective boundary layer (*top*) and above the nocturnal stable boundary layer (below). The disappearance of the frictional force leads to an increase in wind speed and a turning in wind direction (*red arrows*). "L" and "H" indicate minima and maxima in surface pressure and *thin lines* are surface isobars

non-vanishing horizontal synoptic pressure gradients. Therefore, nocturnal low-level jets usually appear at the edges of high-pressure systems (see shaded area in Fig. 3.15).

In the spatial domain this corresponds to a sudden transition of the flow from a surface which is warmer than the air temperature to a smooth surface which is colder than the air temperature. This may happen when the flow crosses the coast line from warm land to a colder ocean surface or from bare land to snow or ice-covered surfaces.

3.4.2.2 Frequency of Low Level Jets

It was mentioned in the preceding section that the occurrence of nocturnal low-level jets depends on certain synoptic weather conditions. Therefore, it can be expected that the frequency of occurrence is linked to the appearance of certain weather or circulation types. For Central Europe the "Grosswetterlagen" (large-scale weather types) have proven to give a good classification of the weather situation (Gerstengarbe et al. 1999). Figure 3.16 shows the frequency of occurrence of low-level jets over Northern Germany as function of these 29 large-scale weather types. The two most relevant types (the two left-most columns in Fig. 3.16) are a high-pressure bridge over Central Europe (type "BM") and a high-pressure area over the British Isles (type "HB"). All in all a low-level jet appeared in 23 % of all nights.

Figure 3.16 showed the frequency of occurrence of a low-level jet as function of the weather type. The relevance of a certain weather type for the formation of a low-level jet can be assessed when comparing the frequency of low-level jet occurrence with the overall frequency of occurrence of the respective weather type. Figure 3.17 has been produced by dividing the frequencies shown in

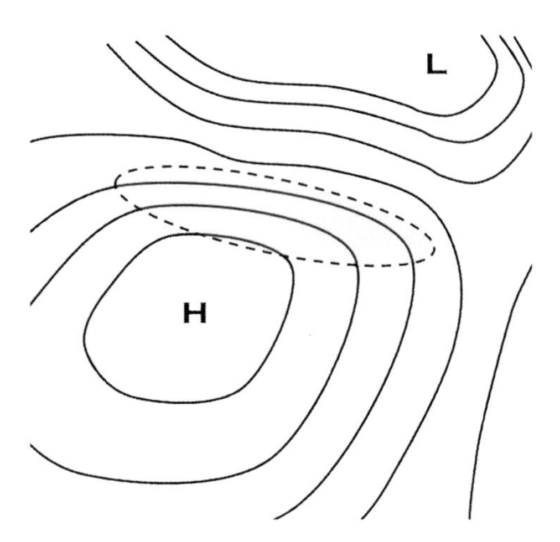

Fig. 3.15 Suitable synoptic conditions for the formation of nocturnal low-level jets are in the shaded area

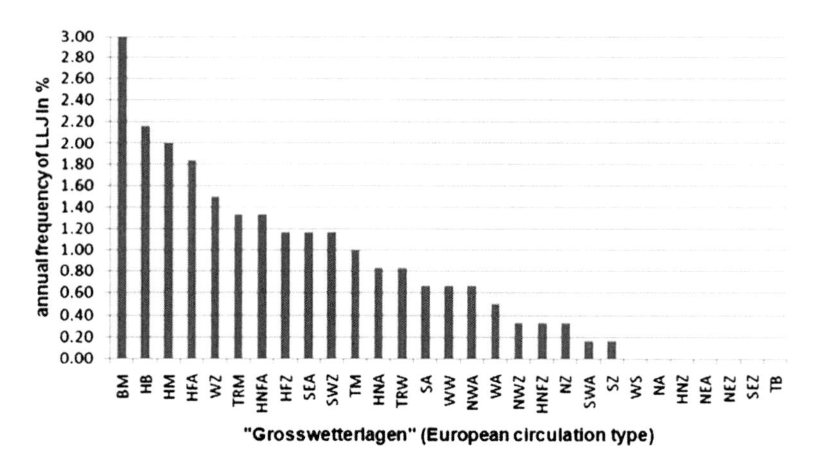

Fig. 3.16 Frequency of the occurrence of nocturnal low-level jets over Northern Germany as function of Central European circulation type (Grosswetterlage, Gerstengarbe et al. 1999) from two years of SODAR data at Hannover (Germany) from autumn 2001 to summer 2003

Fig. 3.16 by the occurrence frequency of the respective weather types during the same observation period. There are two weather types where the occurrence frequency is identical to the occurrence frequency of the low-level jets during this weather type. This means that in every night when this weather type prevailed a low-level jet was observed. This is indicated by a low-level jet efficiency of 1.0 in Fig. 3.17. Small deviations from unity are due to the limited sample size evaluated for this purpose. These two weather types are "HNFA" and "HFZ" which are both related to high-pressure systems to the North of the investigation site.

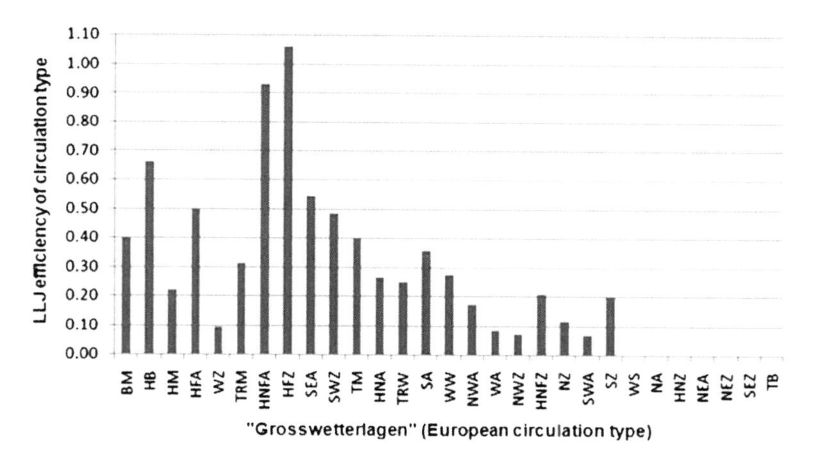

Fig. 3.17 Efficiency of Central European circulation types (Grosswetterlage) to produce a low-level jet from two years of SODAR data

Such a high efficiency for forming a low-level jet allows for a quite certain forecast of the occurrence of a low-level jet. Once such weather types are forecasted a low-level jet will form with a very high probability. The values given in Fig. 3.17 can be used to give the low-level jet formation probability for Northern Germany for each of the weather types. For other areas the investigation has to be repeated with local low-level jet data.

3.4.2.3 Vertical Wind Profiles Below the Jet Core

The vertical wind profile during the occurrence of a low-level jet is modified from (3.16). Empirically we suggest from SODAR measurements:

$$u(z) = u_{\log}(z)\left(1 + \Delta u_{llj}\left(\frac{z_{llj} - z_p - |z_{llj} - z|}{z_{llj} - z_p}\right)^2\right) \quad (3.92)$$

where $u_{log}(z)$ is the equilibrium wind speed from (3.16), Δu_{llj} the relative speed-up with reference to $u_{log}(z)$ in the centre of the low-level jet, z_{llj} the height of the centre of the low-level jet over ground and z_p the height of the surface layer.

A low-level jet is related to a turning of the wind direction with height, $d(z)$ as well. Another empirical relation similar to (3.92) can be formulated:

$$d(z) = d(z_p) + \Delta d_{llj}\left(\frac{z_{llj} - z_p - |z_{llj} - z|}{z_{llj} - z_p}\right) \quad (3.93)$$

where Δd_{llj} is the absolute turning of wind direction from the top of the surface layer to the core of the low-level jet in degrees. This turning may be illustrated

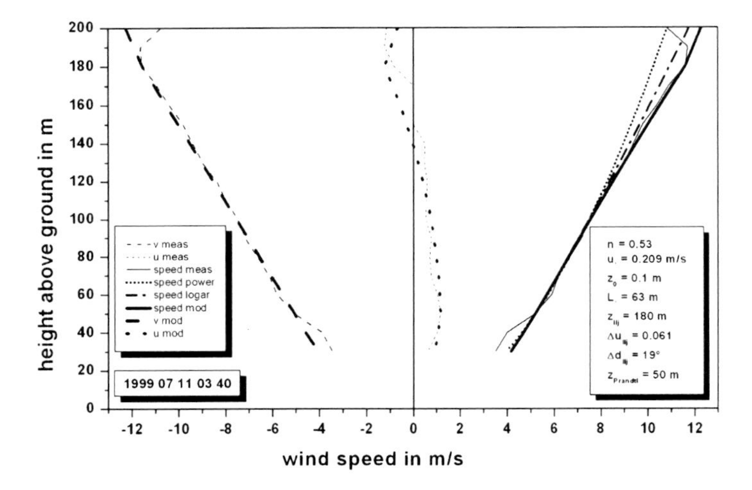

Fig. 3.18 Measured (*thin full lines*) and modelled (*dotted* from (3.22), *dash-dotted* (3.16), *bold* (3.80)) vertical profiles of wind speed and its horizontal components u and v (modelled only) for a rural area. Parameters for (3.16), (3.22) and (3.80) have been chosen in order to have coinciding winds at 50 and 100 m, and are given in the right box. $z_{ref} = 50$ m

with the following two examples from SODAR measurements over rural and urban areas.

Figure 3.18 shows examples for vertical profiles of the west–east wind component u and the south–north wind component v over flat terrain ($z_0 = 0.1$ m), Fig. 3.19 for an urban area ($z_0 = 1$ m). Both Figures demonstrate the ability of (3.92) and (3.93) to describe the vertical turning of the wind underneath a low-level jet.

3.5 Internal Boundary Layers

The boundary layer flow structure over a homogeneous surface tends to be in equilibrium with the surface properties underneath, which govern the vertical turbulent momentum, heat, and moisture fluxes. When the flow transits from one surface type to another with different surface properties, the flow structure has to adapt to the new surface characteristics. This leads to the formation of an internal boundary layer (IBL, internal because it is a process taking place within an existing boundary layer) that grows with the distance from the transition line (Fig. 3.20).

An IBL with a changed dynamical structure can develop when the flow enters an area with a different roughness (e.g. from pasture to forests or from agricultural areas to urban areas) or crosses a coastline. An IBL with a modified thermal structure can come into existence when the flow enters an area with a different surface temperature (e.g. from land to sea or from water to ice). Often dynamical and thermal changes occur simultaneously. Vertical profiles of wind, turbulence,

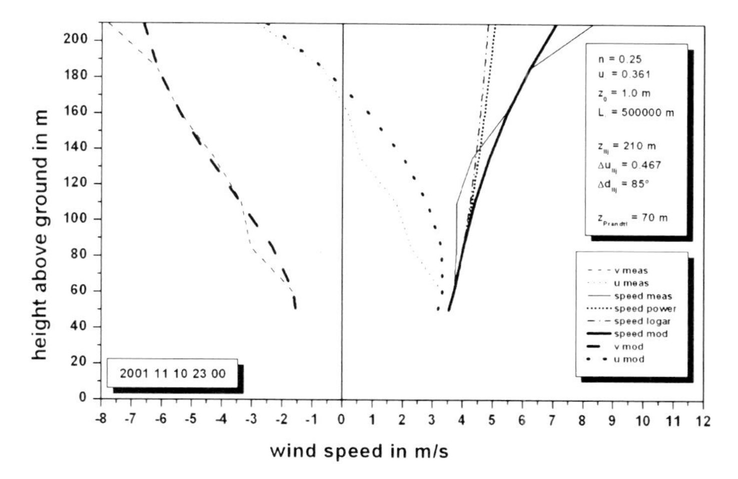

Fig. 3.19 As Fig. 3.18, but for urban terrain ($z_0 = 1$ m)

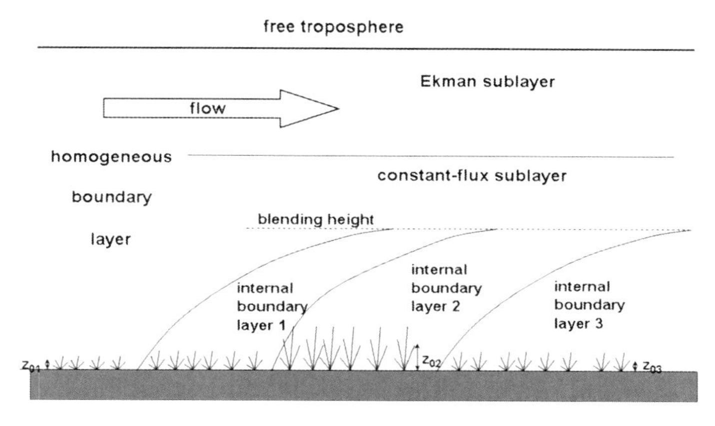

Fig. 3.20 Principal sketch of internal boundary layers developing over step-changes in surface properties

temperature, and moisture are changed within the IBL and return to the undisturbed values at the IBL top.

Slanted IBL tops have to be distinguished from inversions and sloping frontal surfaces at which likewise changes in the vertical profiles of wind, turbulence, temperature, and moisture can happen. Inversions are usually horizontal and caused either by adiabatic sinking motions from above or by radiative cooling from below. Frontal surfaces are slanted like IBL tops but move with the synoptic pressure systems and are not linked to changes in surface properties. If several subsequent changes in surface properties occur in the streamwise direction, multiple IBLs can form. They all grow with distance from the respective boundary of each surface type. At some larger distance to the initiating change in surface

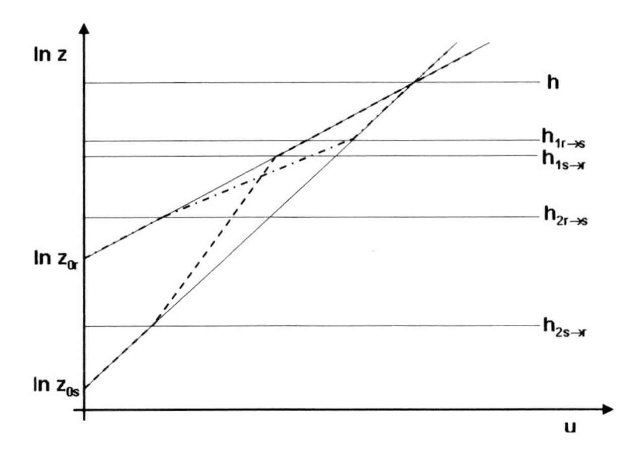

Fig. 3.21 Principal sketch of wind profiles within internal boundary layers developing over step-changes in surface roughness. h denotes the overall IBL height, h_1 the height in the top kink in the wind profile, h_2 the height of the lower equilibrium layer underneath which the flow is in equilibrium with the new surface roughness. s→r means a transition from smooth to rough and r→s a transition from rough to smooth

properties, the single IBL loses its identity and multiple IBLs can no longer be distinguished from each other. This height is also called the blending height.

The height of the top of an IBL is a question of definition (see Fig. 3.21). h_2 gives the height of the layer which is completely in a new equilibrium with the new surface type. This layer is also called the equilibrium layer. h_1 gives the height where the wind profile has its upper kink and matches the undisturbed upstream wind profile. h gives the height in which vertical extrapolations of the near surface equilibrium wind profiles upstream and downstream of the step change meet. The layer between h_2 and h is called the transition layer. The dashed curve in Fig. 3.21 displays schematically the wind profile in the IBL of a smooth-to-rough transition, the dash-dotted line the profile in a rough-to-smooth transition. Real wind profiles show smoother transitions between the vertical layers and exhibit an inflection point between h_2 and h_1.

The description of the height of such IBLs has been the subject of research and data evaluation for several decades now. Savelyev and Taylor (2005) have summarized this work in a review in which they list 20 formulae for the IBL height from earlier publications and add another two. Recently Floors et al. (2011) have revisited the issue having available measurements from upstream and downstream of the change in surface properties. They investigated internal boundary layers forming at the Danish west coast by analysing data from the Horns Rev meteorological tower about 15 km off the coast in the North Sea and the 160 m high Høvsøre onshore mast 1.8 km away from the coastline. Floors et al. found that the dispersion analogy of Miyake (1965) gives the most suitable formula for the IBL height:

$$\frac{h}{z_0}\left(\ln\left(\frac{h}{z_0}\right)-1\right)+1=C\frac{\kappa x}{z_0} \tag{3.94}$$

where x is the distance from the onset of the IBL, z_0 is the maximum of the upstream and downstream roughness, $\kappa = 0.4$ is von Kármán's constant and C is a constant which Floors et al. (2011) put to 2.25. Following Troen and Petersen (1989), the wind profile over the IBL can be written as:

$$u(z)=\begin{cases} u_u \frac{\ln\frac{z}{z_{0u}}}{\ln\frac{c_1 h}{z_{0u}}} & z \geq c_1 h \\ u_d + (u_u - u_d)\frac{\ln\frac{z}{c_2 h}}{\ln\frac{c_1}{c_2}} & c_2 h \leq z \leq c_1 h \\ u_d \frac{\ln\frac{z}{z_{0d}}}{\ln\frac{c_2 h}{z_{0d}}} & z \leq c_2 h \end{cases} \tag{3.95}$$

where z_{0u} is the upstream roughness length, z_{0d} is the downstream roughness length, u_u is the wind speed at the height $c_1 h$ computed from (3.6) using the upstream roughness length and friction velocity, and u_d is the wind speed at the height $c_2 h$ computed from (3.6) using the downstream roughness length and friction velocity. Floors et al. (2011) suggest $c_1 = 0.35$ and $c_2 = 0.07$, because this gives the best fit to their data measured at Horns Rev. The various plots in Savelyev and Taylor show that the height h of the IBL is roughly one tenth of the distance from the step change. This infers that the height of the equilibrium layer is roughly of the order of one hundredth of the distance from the step change. This fits well to the usual rule of thumb which says that a measurement made at a mast at a given height is representative for the surface properties in an upstream distance of about 100 times the measurement height. The advantage of the simple model (3.94) and (3.95) is, that after C, c_1, and c_2 have been specified, only u_{*d}, z_{0u}, and z_{0d} have to be known to describe the wind profile (Floors et al. 2011).

3.6 Wind and Turbulence Profiles Over Forests

In recent years, forests have become an interesting site option for wind turbines since these sites are usually away from larger settlements. Forest-covered surfaces are a special form of vegetated surfaces. The special features of the forest boundary layer decisively depend on the spacing of the trees. If trees grow very close together, their crowns form a rough surface which has much in common with an impervious rough grass land (Raupach 1979) as depicted in Fig. 3.1, having a rather large displacement height [see (3.6)] in the order of two to three thirds of the canopy height. The displacement height substitutes the real Earth's surface in all profile laws for flows over forests. If the trees grow sparser, then the rough surface at the displacement height has to be considered as pervious which is indicated by the bold bended vertical arrow shown in Fig. 3.22. Therefore, the main difference

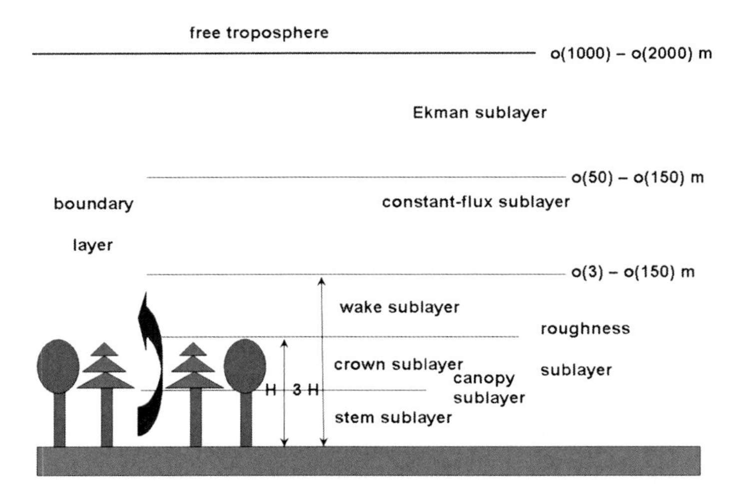

Fig. 3.22 Principal sketch of the vertical structure of the ABL in and over forests

between a densely vegetated forest (Fig. 3.1) and a sparsely vegetated forest (Fig. 3.22) is that larger air parcels can enter (these movements are sometimes called sweeps) and leave (also called ejections) the forest canopy sublayer. This permeability of the rough surface of the forest canopy sublayer leads to an anomaly featuring higher turbulence intensities in the wake sublayer than expected from the mean vertical wind gradient in this layer. Therefore, the usual flux-gradient relationships are not valid in the whole roughness sublayer (see Högström et al. 1989 for details). This anomalous wake layer may extend to about three to five tree heights and has many similarities with an urban surface (see Sect. 3.7 for further details). In contrast to the urban canopy layer, which immerses the entire vertical extend of the buildings, the forest canopy layer must be subdivided into two layers in the vertical: the stem layer and the crown layer. In the stem sublayer, the horizontal wind speed may be higher than in the denser crown sublayer.

Therefore, wind turbines at forest sites should have hub heights of more than about three times the tree height in order to avoid unnecessary fatigue due to enhanced turbulence. Together with the large displacement height which comes with this surface type, this usually means wind turbine hub heights above the displacement height of considerably more than 100 m, i.e. total hub heights of about 130–150 m above ground.

3.7 Winds in Cities

Recent increases in urbanisation have resulted in increased urban energy demands. Research has started to investigate the possibility of local energy generation from wind by turbines especially suitable for urban environments. Such local energy generation would avoid the transport of large amounts of electrical energy from offshore wind farms and desert solar energy plants to the heavily populated cities of the world.

3.7.1 Characteristics of Urban Boundary Layers

Urban agglomerations have recently received special interest in atmospheric boundary layer studies. Nowadays, more than half of mankind is living in cities and the number of megacities with more than 10 million inhabitants is steadily growing. Cities are large pollution sources and because the temperature is already higher than in their surroundings they are especially prone to the effects of a warming climate. All these aspects have fostered studies on the structure of the urban boundary layer (UBL). UBL meteorology has become a special subject in boundary layer meteorology. One aspect of UBL studies is the analysis of wind profiles and thermally driven secondary circulations over cities (urban heat islands). See, e.g., Kanda (2007) and Hidalgo et al. (2008) for an overview of urban meteorology and of urban heat islands. The urban heat island brings about a secondary circulation with winds towards the urban centre near the ground, uprising motion over the urban centre and compensating outflow towards the surrounding rural areas aloft (Shreffler 1978, 1979).

Urban surfaces are characterized by large roughness elements, wide-spread sealed areas, reduced moisture availability at the surface, and increased possibilities for heat storage. This leads to higher turbulence intensities in the urban boundary layer (UBL) and to stronger sensible heat fluxes from the urban surface into the UBL. Both facts induce a greater depth of the boundary layer (see the urban dome in Fig. 3.23). During daytime the reduced moisture availability leads to smaller latent and thus larger sensible heat fluxes at the urban surface compared to rural surfaces. The reduced radiative cooling of the urban surface or even the persisting upward heat fluxes (Velasco et al. 2007) at night prevents the formation of a stable nocturnal boundary layer. Both the increased sensible heat flux during the day and the reduced cooling during the night cause higher temperatures in the UBL compared to the surrounding rural boundary layer. This effect is known as the urban heat island (Atkinson 2003; Chow and Roth 2006). The urban heat island is enhanced by human energy production (Crutzen 2004; Kanda 2007), which with 20–70 Wm^{-2} can be 5–10 % of the energy input by solar irradiation.

In a horizontal flow, the presence of the city results in a change in surface properties. Towns are often isolated islands featuring these special surface properties surrounded by rural terrain so that the flow above them is not in equilibrium with the urban surface. Following Sect. 3.1.1.1, this leads to the formation of internal boundary layers (Fig. 3.23). The internal layer formed by the properties of the urban surface is often called an urban plume.

Following Plate (1995), Roth (2000) and Piringer et al. (2007), the urban boundary-layer (UBL) is usually divided into four layers in the vertical (Fig. 3.24):

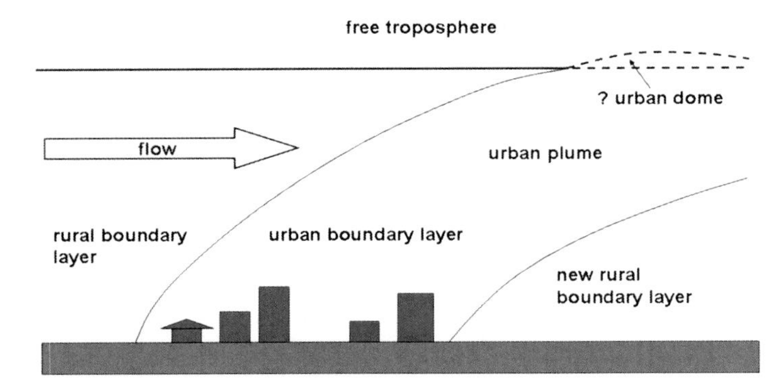

Fig. 3.23 Urban plume downwind of a larger city. This is a special case of an internal boundary layer (cf. Fig. 3.20)

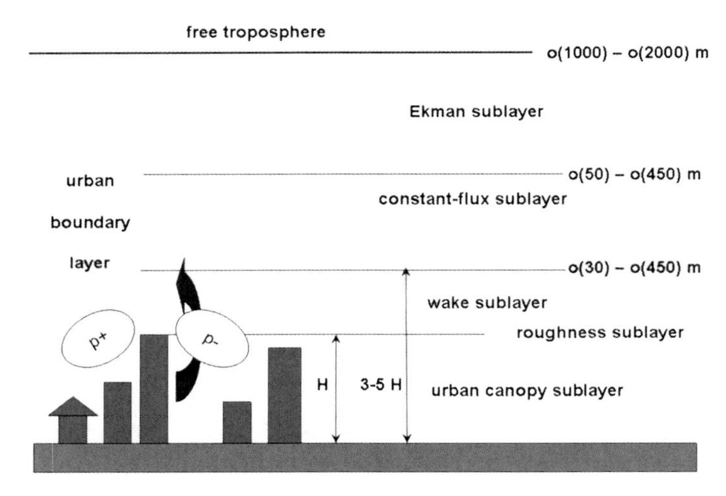

Fig. 3.24 Vertical layering in an urban boundary layer. *H* gives the average building height, $p+$ and $p-$ designate atmospheric pressure disturbances upstream and downstream of single buildings

The lowest one is the urban canopy layer (UCL) which reaches up to the mean top height of the buildings. The next layer is the wake layer in which the influence of single buildings on the flow is still notable. This wake layer usually extends to about two to five times the average building height. These two layers are often jointly addressed as the urban roughness sub-layer (URL, Rotach 1999). Strong vertical exchange by forced vertical motions can occur in this layer. Above the urban roughness layer is the constant flux layer (CFL) or inertial sublayer (IS), which over homogeneous terrain is usually called the surface layer or Prandtl layer. In the uppermost part of the boundary layer above the CFL, the wind direction turns into the direction of the geostrophic wind (the Ekman layer, see

Sect. 3.2). If a convectively driven boundary layer (CBL) is present, no distinction is made between the CFL or Prandtl layer and the Ekman layer but they are jointly addressed as mixed layer. Good overviews of the special features of the UBL can e.g. be found in Roth (2000), Arnfield (2003) and Grimmond (2006).

Wind and turbulence within the UBL are different from flat terrain. Numerous field experiments (for an overview see e.g. Grimmond (2006)), numerical studies (see e.g. Batchvarova and Gryning (2006)) and several wind tunnel studies (Counihan 1973; Farell and Ivengar 1999; Schatzmann and Leitl 2002) therefore have been conducted to investigate the structure of the UBL. Besides of the better understanding of turbulence within the UBL, a realistic representation of the flow field within street canyons and above the buildings is essential for the deployment of suitable urban wind turbines to urban areas (e.g. model simulations for London with ADMS Urban (CERC 2001)).

3.7.2 Vertical Profiles of Wind and Turbulence

Basically, wind profiles over urban areas can be described by the profile laws derived in this chapter above by choosing a large roughness length (usually a metre or more) and a displacement height [see (3.6) and the remarks following this equation] of about two thirds of the mean building height.

Figure 3.25 gives monthly mean wind profiles over a city for four different seasons. The April data in Fig. 3.25 shows the phenomenon of cross-over which has been introduced and explained in Sect. 3.5. The cross-over height is roughly 125 m. This is rather high and probably due to the large aerodynamic roughness of the urban surface that is about 1 m in this case. The August data exhibits the low-level jet phenomenon in the night-time profile at about 325 m above ground even in a monthly average (see Sect. 3.4.1). Both phenomena are closely related and the above given rule of the cross-over height being roughly one third of the height of the low-level jet core is fulfilled as well. The occurrence of these phenomena needs rapid night-time cooling which does not appear over urban heat islands. Therefore, it must be assumed that the low-level jet has formed on a regional scale over the rural environment of this city and has been advected by the mean wind over the city. This again demonstrates the missing horizontal homogeneity for urban boundary layers as depicted in Fig. 3.23.

Figure 3.26 gives an indication on the vertical turbulence profiles over an urban area by showing vertical profiles of the standard deviation of the vertical velocity component from the same measurements as those depicted in Fig. 3.25. Most profiles show an increase with height even for stable stratification. The daytime increase can be explained by unstable stratification [see Eq. (3.18]. The still considerable increase of the night-time values with height evident in the lower 100 m are due to unstable stratification, but above this height they are probably also related to the formation of nocturnal low-level jets (see the upper right and especially the lower left frame in Fig. 3.26). The maximum of this standard

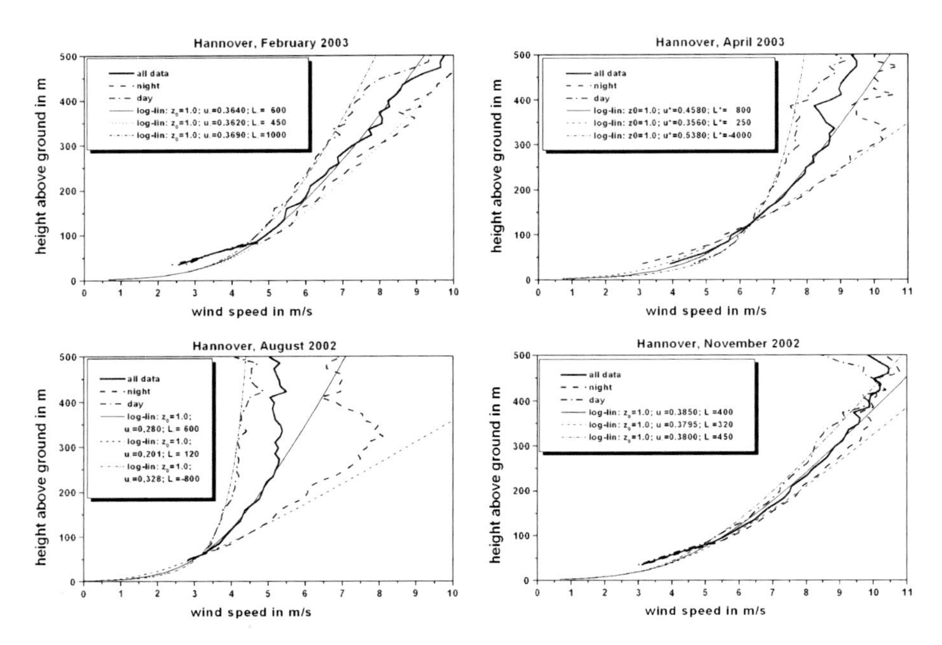

Fig. 3.25 Monthly mean wind profiles over the city of Hannover (Germany) from SODAR measurements (*bold lines*) for February 2003 (*upper left*), April 2003 (*upper right*), August 2002 (*lower left*) and November 2002 (*lower right*). *Thin lines* are computed from Eq. (3.16), the necessary parameters are given in the boxes to the upper left of each frame. *Full lines* show all data, *dash-dotted* lines show daytime data and dashed lines show night-time data

deviation is in the same height as the core of the low-level jet. The ratio of the standard deviation values near the ground to the friction velocity is a little bit higher than expected from Eqs. (3.9) and (3.18) for flat terrain. According to these relations and using the u_* values used for fitting in Fig. 3.25, the standard deviation values should range between about 0.3 m/s for night-time in August and about 0.7 m/s for daytime in April. The differences between daytime and night-time profiles are small in winter and autumn, although the winter profiles show a large dependence of the synoptic wind direction. In wintertime, the largest values of this standard deviation occur with usually stronger westerly winds. Night-time and daytime profiles differ most in spring and summertime. In these seasons, the differences between the mean daytime and night-time profiles are much larger than the differences between the mean profiles for different wind directions.

Figure 3.27 shows monthly mean profiles of the vertical component of the turbulence intensity observed in Hannover, i.e. the standard deviation depicted in Fig. 3.26 divided by the average horizontal wind speed shown in Fig. 3.25. This quantity therefore inversely depends on the mean wind speed. Turbulence intensity is highest in summer and spring. In these two seasons, the daytime values are twice as high as the night-time values. At daytime, turbulence intensity profiles in spring and summer are more or less constant with height up to 300–400 m above ground.

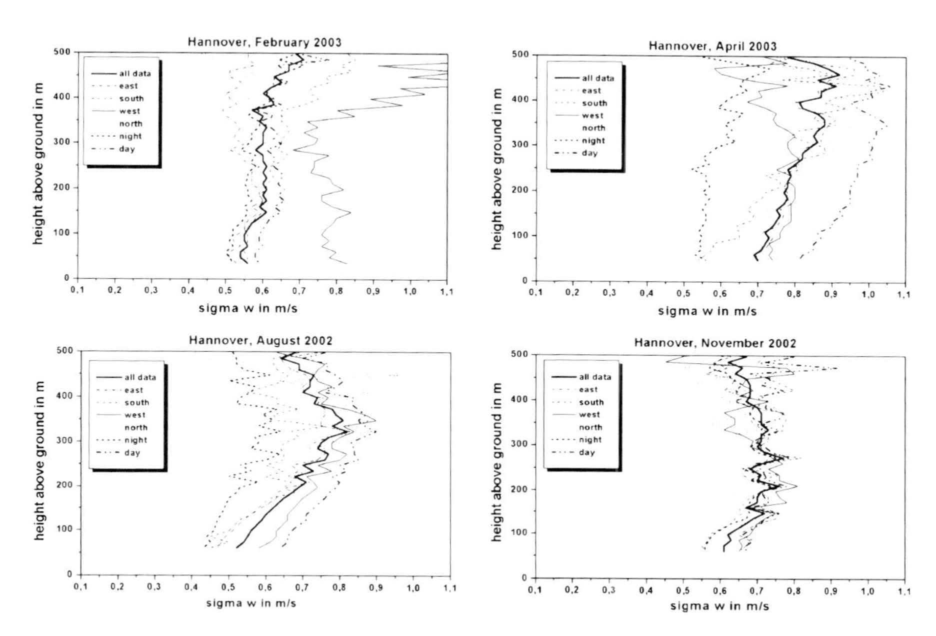

Fig. 3.26 Monthly mean profiles of the standard deviation of the vertical wind component over the city of Hannover (Germany) from SODAR measurements (*bold lines*) for February 2003 (upper *left*), April 2003 (upper *right*), August 2002 (lower *left*) and November 2002 (lower *right*). *Full lines* show all data, *dash-dotted lines* show daytime data and dashed lines show night-time data

In autumn, winter, and generally at night-time, the profiles show a strong decrease of the turbulence intensity with height within the lower 150–200 m.

Similar profiles as those depicted in Figs. 3.25, 3.26, 3.27 have been found over other cities as well (e.g., Moscow in Russia and Linz in Austria, see Emeis et al. 2007b for details). The diurnal course of the variance of the vertical velocity component in summertime is found to be quite similar in Hannover and Moscow. Nevertheless, Fig. 3.28 indicates that the overall level of the standard deviation is somewhat larger over the much larger city of Moscow than over the smaller city of Hannover although the mean wind speeds in Moscow in July 2005 were even lower than in Hanover in August 2002. Both plots show that the standard deviation increases with height at daytime and night-time in summer.

3.7.3 Special Flow Phenomena in Urban Canopy Layers

The flow in the urban canopy layer exhibits special features. Among these are the channelling of flow in street canyons and between taller buildings (see Sect. 4.1), the speed-up of flow over building tops like over hill tops (see Sect. 4.2), the formation of lee-eddies behind buildings, and the high variability of wind

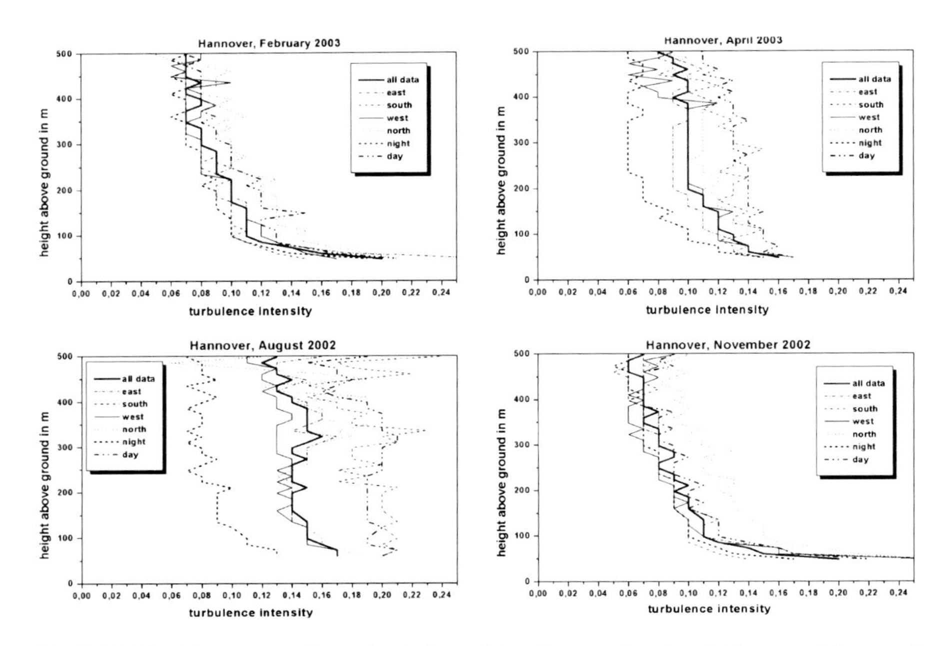

Fig. 3.27 Monthly mean profiles of turbulence intensity over the city of Hannover (Germany) from SODAR measurements (*bold lines*) for February 2003 (upper *left*), April 2003 (upper *right*), August 2002 (lower *left)* and November 2002 (lower *right*). Full lines show all data, dash-*dotted lines* show daytime data and *dashed lines* show night-time data

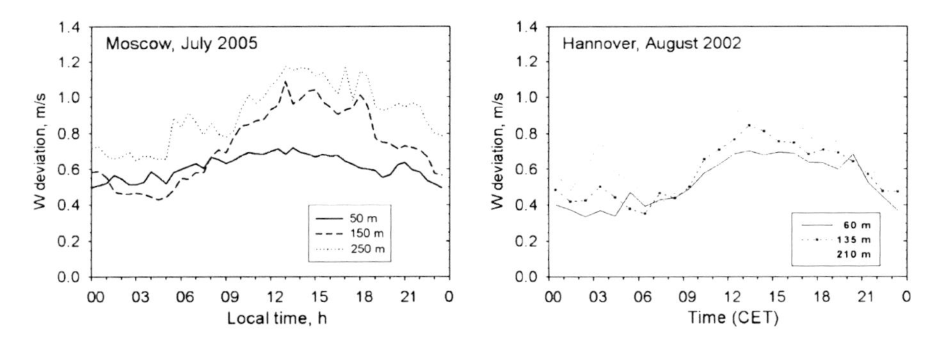

Fig. 3.28 Monthly mean diurnal variation of the standard deviation of the vertical wind component for three different heights for a summer month over Moscow, Russia (*left*) and Hannover, Germany (*right*) plotted against local time

directions. The frequent wind direction changes in urban areas may favour the deployment of smaller turbines with a vertical rotor axis which operate in winds from any direction without adjustment. More details on winds in cities may be found in Cermak et al. (1995).

3.8 Summary for Flat Terrain

Today's wind turbines have hub heights well above the surface layer. Therefore, the wind profiles describing the wind conditions can no longer be based purely on the logarithmic laws (3.6) and (3.16) or the power law (3.22) valid for the surface layer. The profile law of the Ekman layer (3.50) has to be considered for heights above the surface layer. A combined profile for both layers such as those given in (3.65), (3.69) or (3.70) are probably the most suited laws to be used for load assessment and power yield estimates. Equations (3.65), (3.69) and (3.70) apply for the description of the vertical profile of the scale parameter of the Weibull distribution as well.

The diurnal variation of wind speed in the layer above the surface layer is different from the one in the surface layer. Here, in the Ekman layer, in roughly one quarter of the nights in Central Europe, the night-time wind speed is higher than the daytime wind speed. This phenomenon is called low-level jet. The vertical profile of the shape parameter of the Weibull distribution has a maximum at the top of the surface layer due to this phenomenon. Therefore, the relation (3.90) from Justus et al. (1978), which served rather well for the surface layer, is no longer meaningful, but the relation (3.91) from Wieringa (1989) must be used.

Thermal winds (Sect. 2.4) get some relevance in larger heights above ground where the wind shear due to surface friction becomes small. Usually colder air masses coincide with low-pressure areas and warmer air masses with high-pressure areas in temperate latitudes. Therefore, thermal winds usually contribute to an additional increase of wind speed with height.

As really large homogeneous surfaces are rather rare in densely populated areas with frequent land-use changes, the features which come from the development of internal boundary layers described in Sect. 3.5 have to be considered regularly.

For wind turbines erected in forests the special turbulence characteristics addressed in Sect. 3.6 have to be taken into account over pervious forest crown layers. Hub heights should be at least at three times the canopy height in order to avoid enhanced turbulence over pervious forests crown layers. The quite large displacement height deserves special attention for forest sites, because the wind profile laws start from this height and not from the surface.

Typical urban features compared to rural areas are a higher wind shear at heights of several hundreds of metres above ground, a larger increase of turbulence with height especially at night, and a doubling of the turbulence intensity. The nocturnal increase of the standard deviation of the vertical velocity component with height in spring and summer is not just an urban feature but a feature which comes from the interaction between rural and urban air flows. Low-level jets form over rural areas and the additional surface friction due to cities is not sufficient to destroy them. Thus, the higher mechanically-produced turbulence below low-level jets at heights between 100 and 400 m above ground continues the higher thermally-produced turbulence in the urban boundary layer below 100 m. It is obvious that urban areas and forests (see Sect. 3.4) have mechanically some features in common (enhanced turbulence intensity). Looking at thermal features they are very different as forests exhibit no features that are comparable to the urban heat island.

The vertical profiles for urban areas shown in Sect. 3.7.1 are relevant for large wind turbines with hub heights of 100 m and higher. The features discussed in Sect. 3.7.2 are relevant for smaller turbines erected in the urban canopy layer. The numerical modelling of urban boundary layer wind fields needs approaches which go beyond just increasing the surface roughness. Rather, the effects of tall buildings and modified heat and moisture fluxes have to be included as well. An overview of the different approaches using single-layer and multi-layer urban canopy models available today is given in Miao et al. (2009). Only multi-layer models are able to consider the direct influences of taller buildings.

References

Allnoch, N.: Windkraftnutzung im nordwestdeutschen Binnenland: Ein System zur Standortbewertung für Windkraftanlagen. Geographische Kommission für Westfalen, Münster, ARD-EY-Verlag, 160 pp. (1992)

Arnfield A.J.: Two Decades of urban climate research: a review of turbulence, exchanges of energy and water, and the urban heat island. Int. J. Climatol. **23**, 1–26 (2003)

Arya, S.P.: Atmospheric boundary layer and its parameterization. In: Cermak, J.E. et al. (Eds.) Wind Climate in Cities. Kluwer, Dordrecht. 41–66 pp. (1995)

Atkinson B.W.: Numerical modelling of urban heat-island intensity. Bound.-Lay. Meteorol. **109**, 285–310 (2003)

Batchvarova E., Gryning S.-E.: Progress in urban dispersion studies. Theor. Appl. Climatol. **84,** 57–67 (2006)

Blackadar, A.K.: The Vertical Distribution of Wind and Turbulent Exchange in a Neutral Atmosphere. J. Geophys. Res. **67**, 3095–3102 (1962)

Businger, J.A., J.C. Wyngaard, Y. Izumi, E.F. Bradley: Flux profile relationships in the atmospheric surface layer. J. Atmos. Sci. **28**, 181–189 (1971)

CERC: Cambridge Environmental Research Consultants, ADMS dispersion model. (2001) http://www.cerc.co.uk.

Cermak, J.E., A.G. Davenport, E.J. Plate, D.X. Viegas (Eds): Wind Climate in Cities. NATO ASI Series E277, Kluwer Acad Publ, Dordrecht, 772 pp. (1995)

Chow, W.T.L., Roth, M.: Temporal dynamics of the urban heat island of Singapore. Int. J. Climatol. **26**, 2243–2260 (2006)

Counihan, J.: Simulation of an adiabatic urban boundary layerurban boundary layer in a wind tunnel. Atmos. Environ. **7,** 673–689 (1973)

Crutzen, P.J.: New Directions: The growing urban heat and pollution "island" effect – impact on chemistry and climate. Atmos. Environ. **38,** 3539–3540 (2004)

Davis, F.K., H. Newstein: The Variation of Gust Factors with Mean Wind Speed and with Height. J. Appl. Meteor. **7**, 372–378 (1968)

Dyer, A.J.: A review of flux-profile relations. Bound.-Lay. Meteorol. **1** , 363–372 (1974)

Emeis, S., Jahn, C., Münkel, C., Münsterer, C., Schäfer, K.: Multiple atmospheric layering and mixing-layer height in the Inn valley observed by remote sensing. Meteorol. Z. **16**, 415–424 (2007a)

Emeis, S., K. Baumann-Stanzer, M. Piringer, M. Kallistratova, R. Kouznetsov, V. Yushkov: Wind and turbulence in the urban boundary layer – analysis from acoustic remote sensing data and fit to analytical relations. Meteorol. Z. **16**, 393n406 (2007b)

Emeis, S., S. Frandsen: Reduction of Horizontal Wind Speed in a Boundary Layer with Obstacles. Bound.-Lay. Meteorol. **64**, 297–305 (1993)

Emeis, S.: Vertical variation of frequency distributions of wind speed in and above the surface layer observed by sodar. Meteorol. Z. **10**, 141–149 (2001)

Emeis, S.: How well does a Power Law Fit to a Diabatic Boundary-Layer Wind Profile? DEWI-Magazine No. **26**, 59–62. (2005)

Emeis, S.: Vertical wind profiles over an urban area. Meteorol. Z. **13**, 353–359 (2004)

Etling, D.: Theoretische Meteorologie Eine Einführung. 2nd edition, Springer, Berlin, Heidelberg, New York. (2002)

Farell, C., Iyengar, A.K.S.: Experiments on the wind tunnel simulation of atmospheric boundary layers. J. Wind Eng. Indust. Aerodyn. **79**, 11–35 (1999)

Floors, R., S.-E. Gryning, A. Peña, E. Batchvarova: Analysis of diabatic flow modification in the internal boundary layer. Meteorol. Z. **20**, 649–659 (2011)

Foken, T.: Application of Footprint Models for the Fine-Tuning of Wind Power Locations on Inland Areas. DEWI Mag. **40**, 51–54 (2012)

Garratt, J.R.: The atmospheric boundary layer. Cambridge University Press. 334 pp. (1992)

Gerstengarbe, F.-W., P.C. Werner, U. Rüge (Eds.): Katalog der Großwetterlagen Europas (1881 - 1998). Nach Paul Hess und Helmuth Brezowsky. 5th edition. German Meteorological Service, Potsdam/Offenbach a. M. (1999) (available from: http://www.deutscher-wetterdienst.de/lexikon/download.php?file=Grosswetterlage.pdf or http://www.pik-potsdam.de/~uwerner/gwl/gwl.pdf)

Grimmond, C.S.B.: Progress in measuring and observing the urban atmosphere. Theor. Appl. Climatol. **84**, 3–22 (2006)

Gryning, S.-E., Batchvarova, E., Brümmer, B., Jørgensen, H., Larsen, S.: On the extension of the wind profile over homogeneous terrain beyond the surface layer. Bound.-Lay. Meteorol. **124**,251–268 (2007)

Hellmann, G.: Über die Bewegung der Luft in den untersten Schichten der Atmosphäre. Meteorol. Z. **32**, 1–16 (1915)

Hess, G.D., J.R. Garratt: Evaluating models of the neutral, barotropic planetary boundary layer using integral measures. Part I: Overview. Bound.-Lay. Meteor. **104**, 333–358 (2002)

Hidalgo, J., Masson, V., Baklanov, A., Pigeon, G., Gimeno, L.: Advances in Urban Climate Modeling. Ann. N.Y. Acad. Sci. **1146**, 354–374 (2008)

Högström, U., Bergström, H., Smedman, A.-S., Halldin, S., Lindroth, A.: Turbulent exchange above a pine forest, I: Fluxes and gradients. Bound.-Lay. Meteorol. **49**, 197–217 (1989)

Högström, U.: Non-dimensional wind and temperature profiles in the atmospheric surface layer: a re-evaluation. Bound.-Lay. Meteorol., **42**, 55–78 (1988)

Holtslag, A.A.M., H.A.R. de Bruin: Applied modeling of the nighttime surface energy balance over land. J. Appl. Meteor., **27**, 689–704 (1988)

Jensen, N.O.: Change of Surface Roughness and the Planetary Boundary Layer. Quart. J. Roy. Meteorol. Soc. **104**, 351–356 (1978)

Justus, C.G., W.R. Hargraves, A. Mikhail, D. Graber: Methods for Estimating Wind Speed Frequency Distributions. J. Appl. Meteor. **17**, 350–353 (1978)

Kaimal, J.V., J.C. Wyngaard, Y. Izumi, O.R. Coté: Spectral characteristics of surface-layer turbulence. Quart. J. Roy. Meteorol. Soc. **98**, 563n589 (1972)

Kanda, M.: Progress in Urban Meteorology: A Review. J. Meteor. Soc. Jap. **85B**, 363–383 (2007)

Kraus, H.: Grundlagen der Grenzschicht-Meteorologie. Springer, 214 pp. (2008)

Lokoshchenko, M.A., M.A., Yavlyaeva, E.A.: Wind Profiles in Moscow city by the Sodar Data.14th International Symposium for the Advancement of Boundary Layer Remote Sensing. IOP Conf. Series: Earth and Environmental Science 1, 012064. DOI:10.1088/1755-1307/1/1/012064 (2008)

Miao, S., F. Shen, M.A. LeMone, M. Tewari, Q. Li, Y. Wang: An Observational and Modeling Study of Characteristics of Urban Heat Island and Boundary Layer Strutures in Beijing. J. Appl. Meteor. Climatol. **48**, 484–501 (2009)

Miyake, M.: Transformation of the atmospheric boundary layer over inhomogeneous surfaces. Univ. of Washington, Seattle, unpubl. MSc-thesis, Sci. Rep. 5R-6. (1965)

Panofsky, H.A., H. Tennekes, D.H. Lenschow, J.C. Wyngaard: The characteristics of turbulent velocity components in the surface layer under convective conditions. Bound.-Lay. Meteorol., **11**, 355–361 (1977)

Paulson, C.A.: The mathematical representation of wind speed and temperature profiles in the unstable atmospheric surface layer. J. Appl. Meteorol. **9**, 857–861 (1970)

Peña, A., S.-E. Gryning, C. Hasager: Comparing mixing-length models of the diabatic wind profile over homogeneous terrainhomogeneous terrain. Theor. Appl. Climatol., **100**, 325–335 (2010b)

Peña, A., S.-E. Gryning, J. Mann, C.B. Hasager: Length Scales of the Neutral Wind Profile over Homogeneous Terrain. J. Appl. Meteor. Climatol., **49**, 792–806 (2010a)

Peppler, A.: Windmessungen auf dem Eilveser Funkenturm. Beitr. Phys. fr. Atmosph. **9**, 114–129 (1921)

Piringer, M., Joffre, S., Baklanov, A., Christen, A., Deserti, M., de Ridder, K., Emeis, S., Mestayer, P., Tombrou, M., Middleton, D., Baumann-Stanzer, K., Dandou, A., Karppinen, A., Burzynski, J.: The surface energy balance and the mixing height in urban areas – activities and recommendations of COSTAction 715. Bound.-Lay. Meteorol. **124**, 3–24 (2007)

Plate, E.J.: Urban Climates and Urban ClimateModelling: An Introduction. – In: Cermak, J.E. et al. (Eds.) Wind Climate in Cities. NATO ASI Series **E277**, Kluwer Acad Publ, Dordrecht, 23–39. (1995)

Raupach, M.R.: Anomalies in flux-gradient relationships over forest. Bound.-Lay. Meteorol. **16**, 467–486 (1979)

Rotach, M.W.: On the influence of the urban roughness sublayer on Turbulence and dispersion. Atmos. Environ. **33**, 4001–4008 (1999)

Roth, M.: Review of atmospheric turbulence over cities. Quart. J. Roy. Meteor. Soc. **126**, 941–990 (2000)

Sathe, A., S.-E. Gryning, A. Peña: Comparison of the atmospheric stability and wind profiles at two wind farm sites over a long marine fetch in the North Sea. Wind Energy **14**, 767–780 (2011)

Savelyev, S.A., P.A. Taylor: Internal Boundary Layers: I. Height Formulae for Neutral and Diabatic Flows. Bound.-Lay. Meteorol. **115**, 1–25 (2005)

Schatzmann, M., Leitl, B.: Validation and application of obstacle-resolving urban dispersion models. Atmos. Environ. **36,** 4811–4821 (2002)

Schmid, H.P.: Footprint modeling for vegetation atmosphere exchange studies: a review and perspective. Agric. Forest Meteorol. **113**, 159–183 (2002).

Schmid, H.P.: Source areas for scalars and scalar fluxes. Bound.-Lay. Meteorol. **67**, 293–318 (1994)

Schroers, H., H. Lösslein, K. Zilich: Untersuchung der Windstruktur bei Starkwind und Sturm. Meteorol. Rdsch. **42**, 202–212 (1990)

Sedefian, L.: On the vertical extrapolation of mean wind power density. J. Appl. Meteor. **19**, 488–493 (1980)

Shreffler, J.H.: Detection of Centripetal Heat Island Circulations from Tower Data in St. Louis. Bound.-Lay. Meteorol. **15**, 229–242 (1978)

Shreffler, J.H.: Heat Island Convergence in St. Louis during Calm Periods. J. Appl. Meteorol. **18**, 1512–1520 (1979)

Teunissen, H.W.: Structure of mean winds and turbulence in the planetary boundary layer over rural terrain. Bound.-Lay. Meteorol. **19**, 187–221 (1980)

Troen, I., E.L. Petersen: European Wind Atlas. Risø National Laboratory, Roskilde, Denmark. 656 pp. (1989)

Velasco, E., Márquez, C., Bueno, E., Bernabé, R.M., Sánchez, A., Fentanes, O., Wöhrnschimmel, H., Cárdenas, B., Kamilla, A.,Wakamatsu, S., Molina, L.T.: Vertical distribution of ozone and VOCs in the low boundary layer of Mexico City. Atmos. Chem. Phys. Discuss. **7**, 12751–12779 (2007)

Wieringa, J.: Gust factors over open water and built-up country. Bound.-Lay. Meteorol. **3**, 424–441 (1973)

Wieringa, J.: Shapes of annual frequency distributions of wind speed observed on high meteorological masts. Bound.-Lay.Meteorol. **47**, 85–110 (1989)

Zilitinkevich, G.: Resistance Laws and Prediction Equations for the Depth of the Planetary Boundary Layer. J. Atmos. Sci. **32**, 741–752 (1975)

Chapter 4
Winds in Complex Terrain

More and more onshore wind turbines are built away from flat regions near the coasts in complex (i.e., hilly or mountainous) terrain. The most favourite sites in complex terrain are at elevated positions such as hill tops. Therefore, this Chapter introduces a few of the main flow features which influence wind energy yields in complex terrain.

Wind over complex terrain is influenced by changes in surface properties (such as roughness and land use) and the height elevation of the site above sea level (such as hills, ridges, mountains, and escarpments). We will use the term 'topography' to address the whole variation in surface properties and elevation, and we will use the term 'orography' to address especially height elevation. Changes in surface properties without any orographic structures have already been addressed in Sect. 3.5 on internal boundary layers. Here in this Chapter, topographic and purely orographic influences on the wind field will be discussed.

In between roughness and orography we might think of having a third class of topographic features which can be termed flow obstacles, e.g. such as buildings or larger trees (Petersen et al. 1998b). See Sects. 3.6 and 3.7 for a basic treatment of such obstacles.

The complexity of hilly and mountainous terrain does not allow for a straightforward application of the wind profile laws introduced in Chap. 3. Usually, analytical or numerical flow models must be used for the assessment of wind and turbulence conditions at a given site. Three-dimensional numerical wind field models can roughly be stratified into three classes. The simplest ones are mass-consistent flow models which generate a divergence-free flow over orography from given measurements. They do not involve dynamic equations such as (2.1)–(2.4). For reliable solutions they need a larger number of observations. The next class are hydrostatic flow models which solve the dynamic Eqs. 2.2 and 2.3. Equation (2.4) is substituted by the hydrostatic Eq. (2.1). They only work for larger scales of say a few kilometres or more. For smaller scales, full non-hydrostatic models with the full Eqs. (2.2)–(2.4) have to be used.

S. Emeis, *Wind Energy Meteorology*, Green Energy and Technology,
DOI: 10.1007/978-3-642-30523-8_4,

4.1 Characteristics of Boundary Layers Over Complex Terrain

Some basic peculiarities of the boundary layer structure over orographically structured terrain are depicted in Fig. 4.1. Section 4.1.1 and Fig. 4.3 will introduce a major feature of winds in mountainous terrain: the thermally driven mountain and valley winds, and in Sect. 4.1.2 katabatic and drainage winds. Mountain and valley winds as well as katabatic and drainage winds are generated by the orography itself. But there are several other flow features over mountainous terrain, which come from a mainly mechanical modification of the existing larger-scale flow by the underlying orographic features. This includes the acceleration of wind speed in flows passing over hills, mountain tops, ridges and escarpments, the channelling of winds in valleys, gap flows through narrow passages in mountain ranges, the general deflection of winds around single hills and larger mountain ranges. The flow speed-up is described in more detail in Sects. 4.2 and 4.3 below.

Channelling in valleys is a frequent phenomenon that is also visible in wider valleys such as, e.g., in the Upper Rhine valley in Germany. Channelling takes place at least to a height of the accompanying mountain ranges to both sides of such valleys. But often, due to vertical mixing phenomena, channelling extends even above the height of the side ranges. A major feature of channelling is the great constraint which modifies the wind direction distribution. Cross-valley winds only appear rarely. In most of the time we find wind direction along the valley where the selection of one of the possible two directions either depends on the larger-scale pressure field or on the local temperature gradient which constrains the direction of mountain and valley winds passing through a certain location in a valley. The phenomenon of channelling eases the design of larger wind parks, because only two opposite wind directions have to be taken into account in the planning phase. Therefore, siting of the turbines in a wind park in such valleys can be easily optimized. Figure 4.2 gives an example of channelled flow in an Alpine valley in the case of a mountain and valley wind system.

Gap flows occur in a few special locations in a mountain range. The phenomenon is most frequently found in larger mountain ranges perpendicular to the main large-scale wind direction. Gap flows can exhibit quite large wind speeds but are often accompanied by high turbulence as well. As such flows depend decisively on the actual orographic features, no general statements on gap flows can be made here. Gap flows rather need always a specific investigation by on-site measurements with meteorological masts or ground-based remote sensing in order to assess the specific flow features.

An example which combines both the effects of flow channelling in a valley and of a gap flow are the mistral winds in the Rhone valley in Southern France. The river Rhone flows through a major gap between the Massif Central to the West and the French Alps to the East. Mechanisms responsible for the temporal evolution of the Mistral are related to the evolution of upstream synoptic wind speed and direction conditions during the event and the upstream Froude number, calculated in the layer below the upstream inversion height (Caccia et al. 2004).

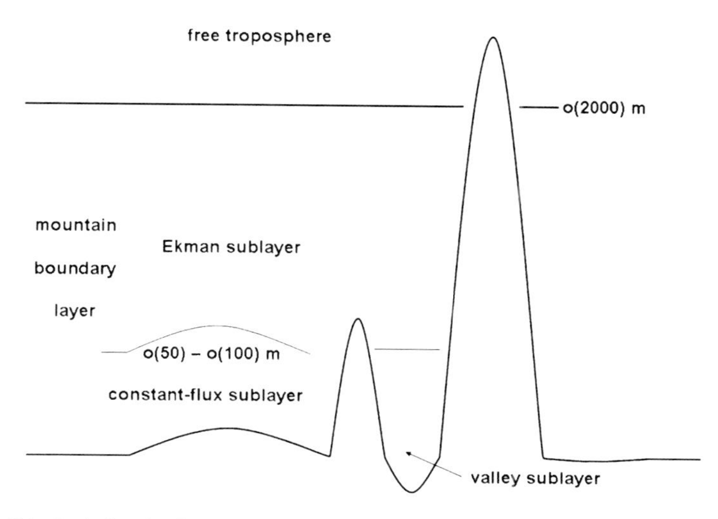

Fig. 4.1 Principal sketch of the vertical structure of boundary layers in mountainous terrain

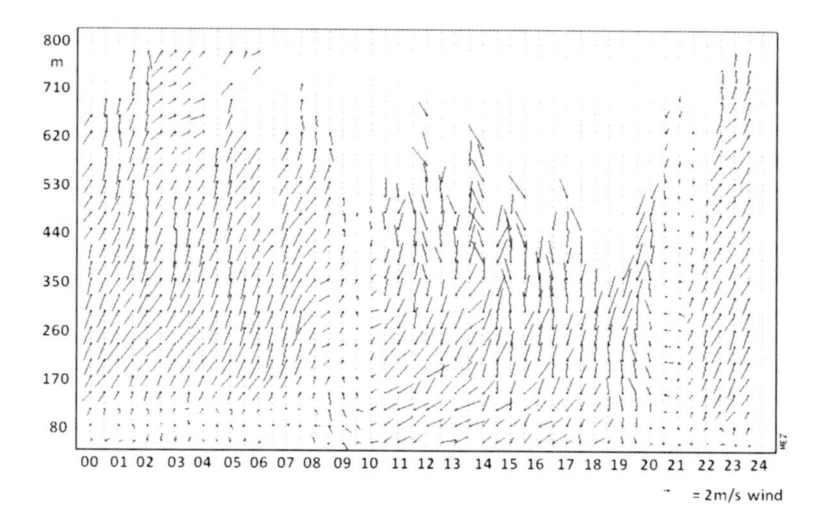

Fig. 4.2 SODAR measurements of the diurnal variation of the horizontal wind field in an Alpine valley demonstrating the diurnal change between mountain winds at night and valley wind at daytime as well as the channelling of the wind in this valley. The wind data have been averaged over 30 min in time and 30 m in the vertical. The abscissa shows 24 h from local midnight to midnight, the vertical axis gives the height in m. Direction of *arrows* gives horizontal wind direction and length of *arrows* gives wind speed

4.1.1 Mountain and Valley Winds

There are local wind systems which do not emerge from large-scale pressure differences but from regional or local differences in thermal properties of the Earth's surface. These local or regional wind systems often exhibit a large regularity so that they can be used for the energy generation from the wind. An overview of such local and regional-scale winds can be found in Atkinson (1981) who presents a wealth of climatological data on this phenomenon.

The presence of hills and mountains leads to much larger horizontal inhomogeneities in the ABL than what was presented in Chap. 3 on homogeneous terrain. Larger mountains can even have a larger vertical extension than the depth of the ABL (Fig. 4.1). Thus, the applicability of the relations given in Chap. 3 above can only be expected to apply in limited parts of the mountain boundary layer, such as over smaller hills or in wide valleys with a flat floor. Differences in the boundary layer over homogeneous terrain come due to both mechanical and thermal forcing. While the mechanical forcing such as channelling of flows in valleys and large-scale blocking by mountain chains are quite obvious, the thermal forcing is more difficult to understand. The thermal forcing is a mixture of the presence of elevated heating (or cooling at night) surfaces and the reduced ratio of the effected air volume to the thermally active surface area in mountainous terrain. As this book concentrates on wind energy generation in the atmospheric boundary layer, aspects gravity wave and foehn generation in thermally stably stratified flows over mountains will not be addressed here. The reader is rather referred to overview papers on these large-scale effects of mountain ranges, e.g. the classical one by Smith (1978) or the book by Atkinson (1981).

Mountains lead to three types of thermally driven secondary circulation systems in the case of weak large-scale pressure gradients and mainly cloud-free skies, which modify the vertical structure of the mountainous ABL: slope winds, mountain and valley winds, and—like land sea wind systems—a diurnally changing system of winds between mountain ranges and the surrounding plains (see Fig. 4.3 which is an extension of the classical sketch from Defant (1949) that depicted only the first two of these three secondary circulation systems). These three phenomena occur on three different spatial scales although all three have the same temporal scale of one day. Slope winds (thin arrows in Fig. 4.3) develop on a slope spatial scale of a few metres up to about 1 km. Mountain and valley winds (full arrows in Fig. 4.3) emerge on a spatial scale of a few hundred metres up to a few 100 km in long valleys. Figure 4.2 displays an example for these intermediate-scale winds. Mountain-plain winds have the largest scale of a few tens of kilometres to more than one hundred kilometres (open arrows in Fig. 4.3). The latter two types of these winds may have some relevance for wind energy generation. The slope winds are probably only interesting for very small wind wheels as the slope wind layer is rather shallow and its depth is varying.

Slope winds come into existence due to the heating by insolation or radiative cooling of a sloping surface in mountainous terrain. These winds emerge and

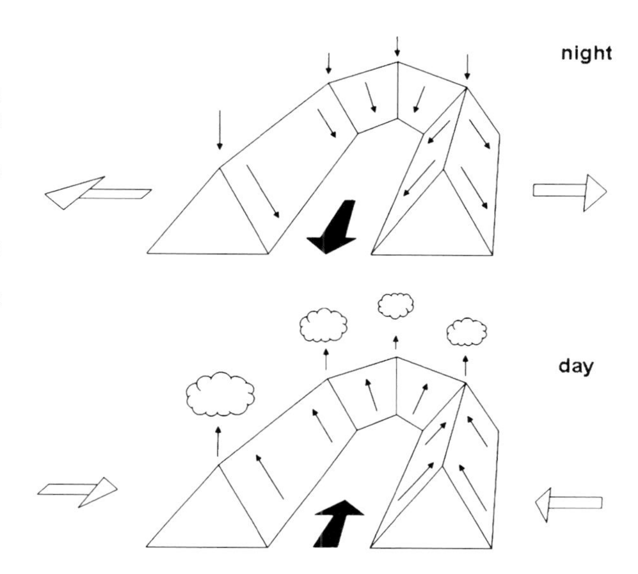

Fig. 4.3 Local and regional wind systems induced by mountains at night-time (*top*) and at daytime (*below*) during calm weather and mainly cloud-free skies. Open arrows denote regional winds towards or away from the mountains, *bold arrows* more local out-valley (*top*) and in-valley (*below*) winds and thin arrows on the mountain flanks indicate purely local slope winds. The thin arrows above the mountains indicate the direction of vertical motion

disappear within minutes with the appearance (and then disappearance) of thermal forcing. They form part of a secondary circulation in a valley cross section. Upslope winds during daytime may lead to compensating sinking motion over the centre of the valley (Vergeiner 1982). This is often the reason why clouds dissolve over the valley centre but form over hill crests. This sinking motion contributes to a stabilization of the thermal stratification in the valley atmosphere and can prolong the existence of temperature inversions in valleys. During the evening, downslope winds develop. See also the description of katabatic winds in Sect. 4.1.2 below.

Mountain and valley winds take a few hours to form. They are a feature of the whole valley (Vergeiner and Dreiseitl 1987). Mountain winds [sometimes called down-valley winds, but a better term would be out-valley winds because local slopes of the valley floor are not decisive (Heimann et al 2007)] usually start 3–4 h after sunset and valley winds (sometimes called up-valley winds or better in-valley winds) 3–4 h after sun rise. Both winds require clear-sky conditions so that heating by incoming short-wave radiation and cooling by outgoing long-wave radiation can occur. The direction of the winds along a valley axis is dominated by the fact that heating and cooling of the valley air is more effective in the narrower upper parts of the valley than in the wider lower parts, because the ratio of air mass to the total thermally active surface is larger in the narrower upper parts of a valley (Steinacker 1984). This differential heating or cooling along the valley axis leads to a pressure gradient along the valley axis which in turn drives the winds. Usually the daytime in-valley winds are stronger and more turbulent than the nocturnal out-valley winds.

The regional-scale wind system between a mountain range and the surrounding planes in Fig. 2.3 has some similarity with the land-sea wind system depicted in

Fig. 5.34 below. This wind system, which blows towards the mountains during daytime and away from the mountains at night-time, takes 4–6 h to develop. It is sometimes observable even 100 km away from the foothills of a larger mountain range (see Lugauer and Winkler 2005 for an example from the European Alps). This wind system comes into existence because at a given height above sea-level the air over the mountains is heated more than over the planes. The opposite occurs at night-time. This differential heating once again leads to a pressure difference at a given height and this pressure difference in turn drives a compensating wind.

There must be a compensating wind system for the mountain and valley winds and for the mountain-plain winds as well. Because this compensating motion takes place over a larger area, it is usually too weak to be differentiated from the synoptic-scale motions. During daytime this compensating motion contributes to downward motions aloft over the surrounding plains of a mountain range that somewhat limits the vertical growth of clouds at the boundary layer top over these plains. For such a circulation system in Southern Germany the term 'Alpine Pumping' has been proposed (Lugauer and Winkler 2005).

4.1.2 Katabatic Winds

Drainage and katabatic flows are purely thermally generated orographic flow features in a mountain boundary layer which have similarity with the above introduced slope winds. They are based on the fact that colder air is heavier than warmer air. Longwave radiative energy losses to space lead to cooling of land, snow and ice surfaces and a compensating downward sensible heat flux, which cools the atmospheric surface layer as well and forms a temperature inversion. In the presence of slopes this induces a horizontal temperature gradient producing a downslope horizontal pressure gradient force (Anderson et al 2005; Renfrew and Anderson 2006) which usually drives shallow drainage flows. These drainage flows are often too shallow in order to be used for wind energy generation.

Drastic examples of deeper drainage flows are the katabatic flows of Antarctica and Greenland. The domed topography and radiative cooling of the snow surface make katabatic flows ubiquitous over these regions (Renfrew and Anderson 2006). Katabatic winds can be very gusty.

4.2 Wind Profiles Over a Hill

Winds over complex terrain show large spatial and temporal variations. Nevertheless there exist a few analytical approaches that help to analyse at least first-order features of attached flow over complex terrain. Non-linearity such as flow separation cannot adequately be described with analytical models but must be addressed with non-linear numerical flow models, see e.g., Zenman and Jensen

(1987). Some of the linear approaches are quite old and date back to work of, e.g., Jackson and Hunt (1975). These analytical approaches have always been accompanied by numerical efforts, see e.g., the work of Taylor (1977). Also the well-known WAsP model is based on such linear analytical approaches (Troen and Petersen 1989).

4.2.1 Potential Flow

The simplest case for a description of flow over a hill is frictionless potential flow. This implies a laminar flow of a non-viscous fluid with no surface friction. It is presented here in order to present an analytical model that shows first-order effects of flow over hills. The main feature is the speed-up of the wind speed over the hill, a slight wind speed reduction upstream of the hill and a considerable reduction of the wind speed over the downwind slope of the hill.

For a flow perpendicular to a two-dimensional ridge (i.e. a ridge which is infinitely long in the direction perpendicular to the flow), the speed-up of the potential flow over the hill can be described using the thin airfoil theory (Hoff 1987):

$$\Delta u_{pot}(x,z) = u_{\infty}(L)\frac{H}{L}\sigma\left(\frac{x}{L},\frac{z}{L}\right) \tag{4.1}$$

where x is the direction perpendicular to the ridge, z is the vertical coordinate, H is the height of the ridge, L is the half-width of the ridge (the distance from the crest to the place where the height is $H/2$), $u_{\infty}(L)$ is the scaling wind speed in the undisturbed flow at height L. Therefore, all heights in this simple model scale with L. σ is the form function of the ridge cross-section. H/L is the aspect ratio of the ridge and describes the magnitude of the slope. Adding (4.1) to the undisturbed flow, $u_{\infty}(z)$ yields for the wind profile in the potential flow over the ridge:

$$u_{pot}(x,z) = u_{\infty}(z) + u_{\infty}(L)\frac{H}{L}\sigma\left(\frac{x}{L},\frac{z}{L}\right) \tag{4.2}$$

In contrast to all wind profile relations given in Chap. 3, the wind profile relation (4.2) does not only depend on the vertical coordinate but also contains a horizontal coordinate. The form function σ can be given analytically as long as the ridge cross-section h(x) can be described by the inverse polynom (see Fig. 4.4):

$$h\left(\frac{x}{L}\right) = \frac{1}{1+\left(\frac{x}{L}\right)^2} \tag{4.3}$$

The associated form function σ for this ridge cross-section (4.3) reads:

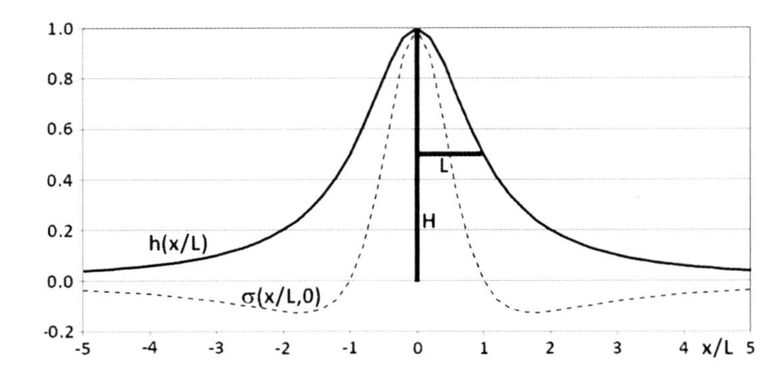

Fig. 4.4 Ridge function $h(x/L)$ (*full line*) and form function $\sigma(x/L, 0)$ (*dashed line*). Ridge height H and half-width L are indicated as well

$$\sigma\left(\frac{x}{L},\frac{z}{L}\right) = \frac{\left(1+\left(\frac{z}{L}\right)\right)^2 - \left(\frac{x}{L}\right)^2}{\left(\left(1+\left(\frac{z}{L}\right)\right)^2 + \left(\frac{x}{L}\right)^2\right)^2} \tag{4.4}$$

For the position of the ridge crest ($x = 0$) we obtain the following special relation:

$$\sigma\left(0,\frac{z}{L}\right) = \frac{1}{\left(1+\left(\frac{z}{L}\right)\right)^2} \tag{4.5}$$

Equation 4.5 describes the decrease of the form function with height that is a function of the half-width of the ridge only. The wider the ridge the higher up the hill influences the flow. The vertical profile of the potential flow speed over the ridge crest is thus:

$$u_{pot}(0,z) = u_\infty(z) + u_\infty(L)\frac{H}{L}\frac{1}{\left(1+\left(\frac{z}{L}\right)\right)^2} \tag{4.6}$$

This vertical wind profile function (4.6) is unrealistic when approaching the surface, because potential flow is without friction and therefore the flow speed in potential flow does not vanish at the surface. Rather, the contrary is the case and the potential flow speed is at its maximum at the ridge crest. There we have ($x = 0$, $z = 0$) $\sigma = 1$ and

$$u_{pot}(0,0) = u_\infty(0) + u_\infty(L)\frac{H}{L} \tag{4.7}$$

Equation 4.7 means that the speed-up of the wind speed over a ridge crest is proportional to the slope of the flanks of the ridge. The form function (4.4) cannot be given analytically for a Gaussian-shaped hill. Numerical integration yields a slightly lower value than for the function given in (4.4) with $\sigma_{Gauss}(0,0) = 0.939$.

This implies that the relative speed-up $\Delta u/u_\infty$ over a 100 m high ridge with a half-width of 1,000 m is about 10 % or:

$$\frac{\Delta u}{u_\infty} \approx \frac{H}{L} \tag{4.8}$$

4.2.2 *Modifications to the Potential Flow: Addition of an Inner Layer*

As said above, the potential flow solution is unrealistic when approaching the surface, because it produces a solution which is symmetrical to the crest line. The potential flow solution is valid in an outer layer only. The decrease of the wind speed towards zero speed at the surface (non-slip condition) takes place in an inner layer with depth l within which the surface friction dominates. This has led to the idea of a two-layer model (Jackson and Hunt 1975). The depth of the inner layer depends on the half-width L again. Jackson and Hunt (1975) derived the following implicit relation for l:

$$l \ln\left(\frac{l}{z_0}\right) = 2\kappa^2 L \tag{4.9a}$$

with the surface roughness length z_0. Jensen et al (1984), Mason (1986), and Hoff (1987) derived a similar but slightly different relation:

$$l \ln^2\left(\frac{l}{z_0}\right) = 2\kappa^2 L \tag{4.9b}$$

For large values of L/z_0, the height of the inner layer calculated from Eq. 4.9b is much smaller than calculated from (4.9a) (see Fig. 4.5). Roughly spoken, the inner layer depth from (4.9a) is of the order of 3–6 % of the half-width of the ridge (Fig. 4.3 right), or—from Eq. (4.9b)—in the order of 1–2 % of the half-width of the ridge. Experimental data from Taylor et al. (1987) and Frank et al. (1993) support the latter formulation (4.9b).

As stated above after Eq. (4.6) the potential flow solution is unrealistic when directly approaching the surface. The true wind profile can be described by matching the potential flow profile (4.2) for the outer layer above l with the logarithmic profile (3.6) for the inner layer:

$$u(x, z < l) = u_\infty(z) + u_\infty(z)\frac{\ln\frac{L}{z_0}}{\ln\frac{l}{z_0}}\frac{H}{L}\sigma\left(\frac{x}{L}, \frac{z}{L}\right) = u_\infty(z) + \Delta u(x, z < l) \tag{4.10}$$

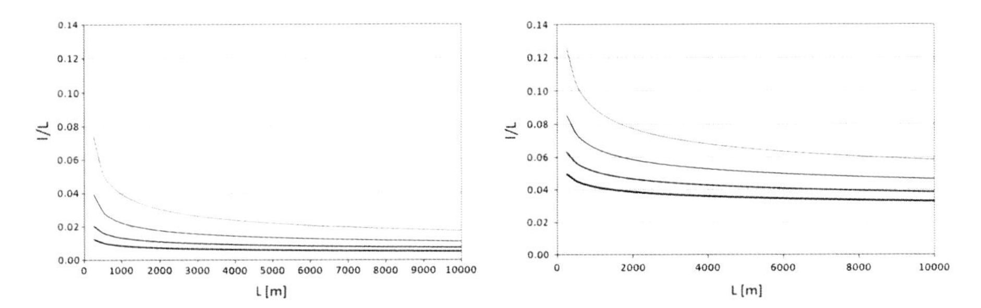

Fig. 4.5 Depth of the inner layer, l as function of the half-width, L and the surface roughness length z_0, found as an iterative solution of (4.9b) (*left*) or (4.9a) (*right*). Lowest curve: $z_0 = 0.02$ m, second curve: $z_0 = 0.1$ m, third curve: $z_0 = 0.5$ m and *upper* curve: $z_0 = 2.5$ m

Equation (4.10) fulfills the non-slip condition at the surface. Hoff (1987) gives the following relation which considers also the surface pressure gradient across the ridge by an additional term:

$$u(x, z < l) = u_\infty(z) + \Delta u(x, z < l) + \delta u(x, z < l) \tag{4.11}$$

with the pressure gradient-related term:

$$\delta u(x, z < l) = \frac{1}{\kappa} \delta u_* \left(\frac{x}{L}\right) \ln\left(\frac{z}{z_0}\right) \tag{4.12}$$

which requires a modified formulation for the friction velocity:

$$\delta u_* \left(\frac{x}{L}\right) = -\frac{l}{2\rho u_{*\infty}} \frac{\partial p}{\partial x} = u_{*\infty} \frac{\ln\left(\frac{L}{z_0}\right)}{\ln\left(\frac{l}{z_0}\right)} \frac{H}{L} \Delta\sigma\left(\frac{x}{L}\right) \tag{4.13}$$

The increment $\Delta\sigma$ in Eq. (4.13) is the horizontal difference of the form function σ in the range between $x/L—D$ and $x/L + D$, where D is supposed to be small compared to L:

$$\Delta\sigma\left(\frac{x}{L}\right) = \frac{1}{2D}\left(\sigma\left(\frac{x}{L} + D, \frac{z}{L} = 0\right) - \sigma\left(\frac{x}{L} - D, \frac{z}{L} = 0\right)\right) \tag{4.14}$$

Smooth vertical wind profile functions which cover both the inner and the outer layer can be formulated as follows (Hoff 1987):

$$u(x, z) = u_\infty(z) + u_\infty(L)\frac{H}{L}\sigma\left(\frac{x}{L}, \frac{z}{L}\right)P_0(z) + \frac{1}{\kappa}\delta u_*\left(\frac{x}{L}\right)\ln\left(\frac{l}{z_0}\right)P_\delta(z) \tag{4.15}$$

with:

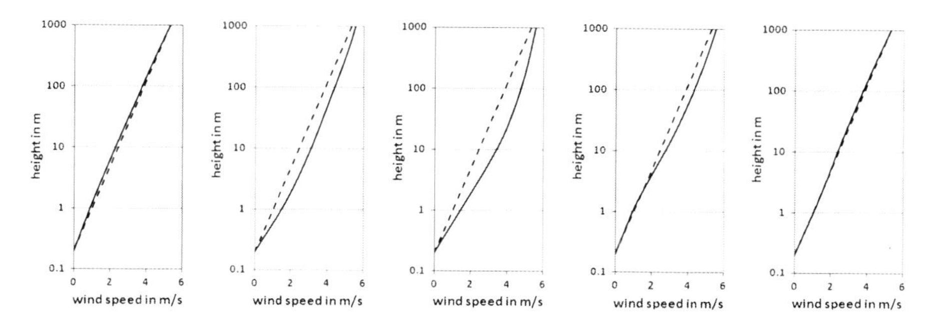

Fig. 4.6 Vertical wind profiles over the ridge shown in Fig. 4.2 for $L = 1{,}000$ m, $H = 200$ m, $z_0 = 0.2$ m and $u_{*\infty} = 0.25$ m/s at $x/L = -2$, -0.5, 0 (*crest line*), 0.5 and 2. *Full line* from Eq. 4.15, *dashed line* for horizontally flat terrain

$$P_0(z) = 1 + \frac{\ln\left(\frac{z}{l}\right)}{\ln\left(\frac{l}{z_0}\right)} \exp\left(-\frac{z - z_0}{l}\right) \tag{4.16}$$

and:

$$P_\delta(z) = \frac{\ln\left(\frac{z}{z_0}\right)}{\ln\left(\frac{l}{z_0}\right)} \exp\left(-2\left(\frac{\ln\left(\frac{z}{z_0}\right)}{\ln\left(\frac{l}{z_0}\right)}\right)^2\right) \tag{4.17}$$

Figure 4.6 shows sample results from (4.15) using (4.9b) for a ridge with half-width $L = 1{,}000$ m and aspect ratio $H/L = 0.2$. $x/L = -2$ is upstream of the ridge closely before the minimum of the shape function σ (see Fig. 4.2). $x/L = -0.5$ and 0.5 are at the positions where the shape function σ has its largest gradients. $x/L = 0$ is on the crest of the ridge and $x/L = 2$ has been chosen symmetrically to the first point. We see the largest speed-up over the crest itself at the top of the inner layer at the height of the length l which was at roughly 16.5 m above ground in this example (see Fig. 4.3). The vertical wind shear is enhanced below this height l compared to the undisturbed logarithmic profile (dashed line) and the shear is reduced above this height. The two frames to the right show the influence of the wake. This influence leads to a reduced wind speed near the height l, although this analytical model is not able to produce flow separation which should set in for aspect ratios larger than about 0.2.

In the outer layer, the solution is still symmetrical to the hill crest, but in the inner layer a considerable asymmetry becomes visible. In this respect, solution (4.15) is more realistic than the pure potential flow solution in Sect 4.2.1. Nevertheless it has to be noted that the analytical model (4.15) can only be used for shallow hills with aspect ratios smaller than 0.2 and a cross-wind elongation which is much larger than the width of the ridge cross-section parallel to the wind direction. The atmospheric stability in this analytical approach is limited to neutral conditions.

A different approach which divides the flow field in three layers has been developed by Sykes (1980). He distinguished the following layers: a very thin wall layer, a Reynolds-stress sublayer across which the Reynolds stresses vary rapidly, and an outer layer. The flow perturbations due to the presence of the hill are calculated for different orders of the slope $\varepsilon^{1/2} = H/L$ ($\varepsilon \ll 1$). The height of the Reynolds-stress sublayer is of the order εL. For an aspect ratio of $H/L = 0.1$ this is quite close to the inner layer height from (4.9b).

4.2.3 Modifications to the Potential Flow: Consideration of Thermal Stability

As a preparation we rewrite Eq. (4.1) in terms of the fractional speed-up:

$$\Delta s(x,z) = \frac{\Delta u_{pot}(x,z)}{u_\infty(l)} = \frac{u_\infty(L)}{u_\infty(l)}\frac{H}{L}\sigma\left(\frac{x}{L},\frac{z}{L}\right) \tag{4.18}$$

Bradley (1983) studied the dependence of the fractional speedup ratio on stability. As a first approximation, Bradley assumed that Eq. (4.18) is still valid for non-neutral flow as long as buoyancy forces are small compared to pressure gradient forces. Then (4.18) is approximately valid but the velocities $u_\infty(L)$ and $u_\infty(l)$ are calculated from diabatic Monin–Obukhov velocity profiles (3.16). For non-neutral stratification, one obtains:

$$\Delta s(x,z) = \frac{\ln\frac{L}{z_0} - \Psi\left(\frac{L}{L_*}\right)}{\ln\frac{l}{z_0} - \Psi\left(\frac{l}{L_*}\right)}\frac{H}{L}\sigma\left(\frac{x}{L},\frac{z}{L}\right) \tag{4.19}$$

where L_* designates the Obukhov length [see (3.11)]. The stability function Ψ is given in (3.15) and (3.21). Ψ has been limited to a minimum value of -5 according to Eq. (31) in Frank et al. (1993). Δs increases with increasing stability and is reduced with unstable flow (Fig. 4.7). This becomes also intuitively clear, because increasing stability opposes to the vertical displacement of the streamlines over the hill. Thus, the streamlines are squeezed together and the speed-up is increased. Evidence from real data is depicted, e.g., in Fig. 2 of Frank et al. (1993).

4.2.4 Weibull Parameters over a Hill

SODAR measurements on a hill top have been evaluated in Emeis (2001) to derive vertical profiles of the two Weibull parameters over a hill. The form parameter is

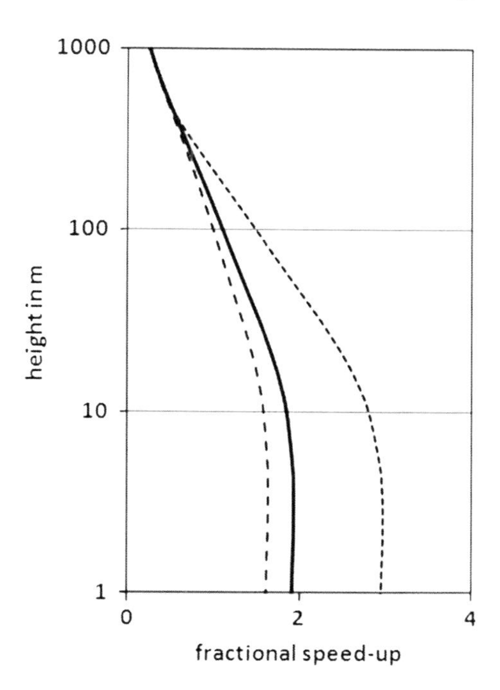

Fig. 4.7 Stability dependent fractional speed-up from Eq. 4.19. *Full line*: neutral stratification, *dashed line* unstable ($L_* = -500$ m), *short-dashed line* stable ($L_* = 500$ m)

described by an analogue to Eq. (3.54), the shape parameter is described by (3.91). Figure 4.8 gives examples from SODAR measurements (see Emeis 2001 for details).

The fact that the vertical profile of the scale parameter is much better described by the simplified Ekman law (3.54) using $\gamma = 0.035$ instead of by the surface layer profiles (3.6) or (3.22) indicates that the wind profile over a hill and hence the vertical profile of the scale parameter behaves like vertical wind profiles in the Ekman layer. This is understandable since the hill top reaches above the surface layer into the Ekman layer. The parameters $z_m = 50$ m and $c_2 = 0.01$ have been used to produce the curve which fits to the October curve in Fig. 4.8 right. For a fit to the September and November curves a value of $c_2 = 0.03$ would be more appropriate. Again, the profiles from Justus et al. (1978) and Allnoch (1992) [see Eq. (3.90)] do not fit to reality.

4.3 Wind Profiles Over an Escarpment

A slightly more complex flow is the flow over an escarpment which gathers features of the upstream side of a hill (Sect. 4.2) and of an internal boundary layer (Sect. 3.5). See Fig. 4.9 for a principal sketch.

In case of an isolated hill, the flow returns to its original state somewhere behind the obstacle. In case of a roughness change, an internal boundary layer forms which finally replaces the old boundary layer. The flow over an escarpment

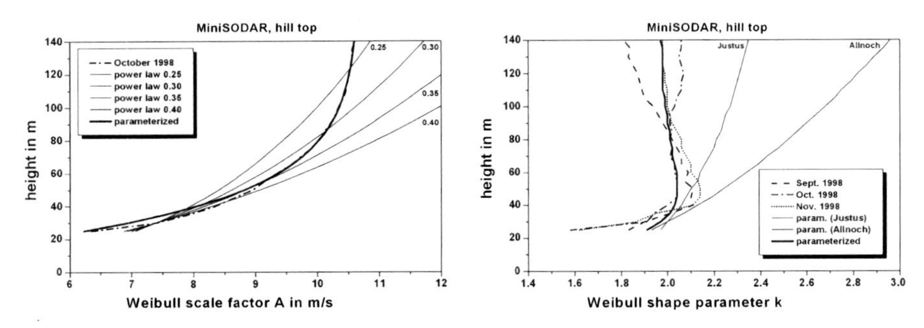

Fig. 4.8 As Fig. 3.12. Weibull parameter over a hill from SODAR measurements compared to analytical profiles. *Left* scale parameter, *right* form parameter. The parametrized curve in the *left* frame is obtained from (3.54) putting $A_g = 10.67$ m/s and $\gamma = 0.035$, the one in the *right* frame from (3.91) putting $z_m = 50$ m and $c_2 = 0.01$. The curves labelled "Justus" and "Allnoch" have been computed from (3.90)

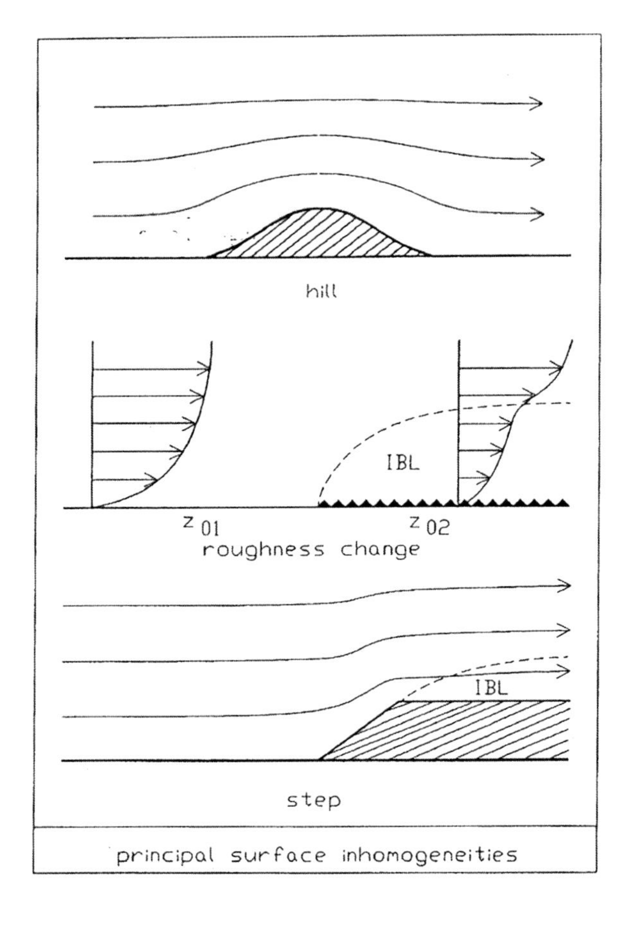

Fig. 4.9 Principal outline of a flow over an escarpment (from Emeis et al 1995)

should be in some way a mixture between the flow over a hill and that over a roughness change. Experimental data on flow over gentle escarpments were obtained by Bowen and Lindley (1977) and Bowen (1979), which Astley (1977) compared with numerical calculations. A comparison between an experimental study and analytical approaches to this flow problem can be found in Jensen (1983).

Figure 4.10 shows some sample data for flow over an escarpment from measurements at Hjardemål at the Danish west coast (Emeis et al. 1995). The escarpment is about 16 m high and the slope of the escarpment is about 30 m wide. This leads to a mean aspect ratio H/L of about 0.5 or 28°. Here, H denotes the height and L the width of the escarpment slope. The measurement line was perpendicular to the escarpment from 400 m upstream to 300 m downstream. Main measurement heights were 5 and 10 m above the surface. Additional instruments were mounted at some sites between 2 and 24 m. Mean wind speed and wind fluctuations were measured with cup anemometers and ultrasonic anemometers. Figure 4.10 shows the speed-up Δs, the longitudinal standard deviation σ_u (parallel to the local surface), the vertical standard deviation σ_w (perpendicular to the local surface) and the friction velocity u_* from 50 m upstream of the upper edge of the escarpment to 50 m downstream. The standard deviations and the friction velocity are normalised with their upstream values at 400 m upstream of the escarpment.

All four frames in Fig. 4.10 show data for slightly unstable thermal stratification in top position ($-0.05 < z/L_* < 0$), near neutral conditions in the middle ($0 < z/L_* < 0.18$) and stable conditions below ($0.18 < z/L_* < 0.29$). The upper left frame of Fig. 4.10 shows the increase of the speed-up with increasing thermal stability of the flow. Additionally, an area with reduced flow speeds is discernible upstream of the escarpment. This flow reduction becomes more pronounced with stable stratification. The maximum speed-up of 62 % with neutral stratification at $z/H = 0.125$ fits well to values from wind tunnel experiments given by Bowen and Lindley (1977), who found a speed-up of 70 % at $z/H = 0.2$. Variations of the slope of the escarpment in Bowen and Lindley's experiment showed that the maximum speed-up for larger slopes no longer depended on the slope.

While σ_u and σ_v (not shown) are relatively little influenced by the escarpment, σ_w shows a strong reaction to the presence of the escarpment. σ_u only exhibits changes of more than 10 % for neutral and unstable stratification downstream of the upper edge of the escarpment in agreement with the wind tunnel data of Bowen and Lindley (1977). σ_v shows slight stability dependence over the slope (increasing with increasing stability). σ_w has a maximum increase of 55–70 % shortly upstream of the upper edge of the escarpment.

The inner layer (see Sect. 4.2.2) was not captured in this experiment, because this would have required measurement at heights lower than roughly 0.16 m which was technically not feasible. The flow in the outer layer can be approximately described by the Eqs. (4.1–4.4) as well as it was possible for the flow over a ridge. Moreover, the function $\sigma(x/L,z/L)$ in (4.4) cannot be given analytically but must be determined numerically.

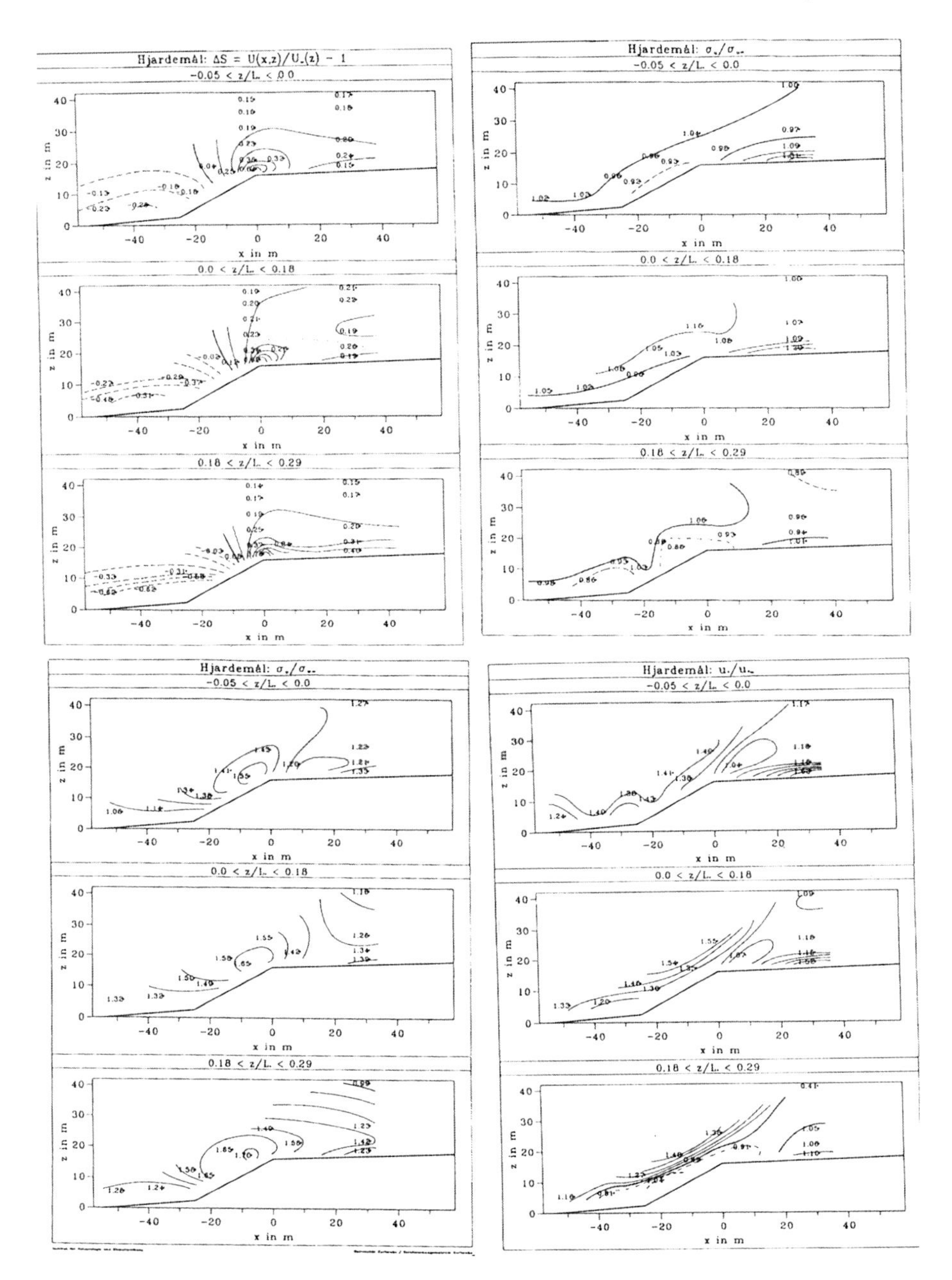

Fig. 4.10 Fractional speed-up (*upper left*), normalised longitudinal standard deviation (*upper right*), normalised vertical standard deviation (*lower left*) and normailsed friction velocity (*lower right*) from ultrasonic anemometer measurements at an escarpment in Denmark (from Emeis et al. 1995). Normalisation was made with the respective undisturbed values observed 400 m upstream

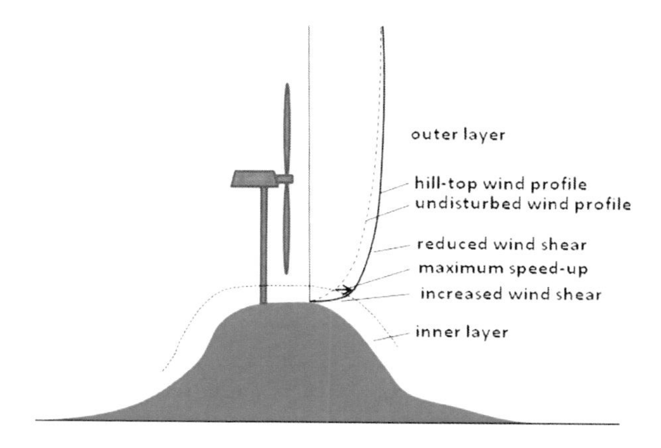

Fig. 4.11 Principal sketch of wind conditions over a ridge. Modern wind turbines are usually in the outer layer of the ridge-crossing flow

4.4 Spectra

The frequency dependence of the power of turbulent fluctuations is described by turbulence spectra as introduced in Sect. 3.3. Special turbulence spectra over complex terrain are given in Panofsky et al. (1982) and Founda et al (1997). Founda et al (1997) found good agreement between measurements over a hill top and the spectra given in Eqs. (3.78)–(3.80), because it turned out to be difficult to determine an appropriate value for the friction velocity. Founda et al (1997) used $L_i^x = 1/(2\Pi)/k_i^p$ instead of (3.84).

4.5 Diurnal Variation

The diurnal variation of the wind speed over ridges and mountain crests resembles the diurnal variation in the Ekman layer, because these crests are usually above the surface layer at night-time. Hills and lower mountains which are smaller than the boundary layer height may develop a shallow surface layer over them as long as they are quite smooth. Due to the change between boundary layer conditions at daytime and free-atmosphere conditions at night-time, wind speeds over crests are usually higher at night-time than at daytime.

4.6 Summary for Complex Terrain

The main peculiarity of flow over a hill or a mountain chain is the speed-up of the wind speed over the summit or the crest line. The boundary layer over the crest can be separated into two layers. There is a rather thin inner layer within which frictional forces dominate over inertial forces. This layer has a depth of typically

1–2 % of the half-width of the hill or mountain chain. Above the inner layer is the outer layer within which the inertial forces dominate. The fractional speed-up is at maximum at the boundary between the inner and the outer layer. Modern wind turbines with hub heights around 100 m and more are usually in the outer layer (Fig. 4.11). Therefore, they are exposed to less vertical wind gradients than over level terrain. Building even higher towers with larger hub heights thus gives only a relatively low gain in power yields.

Section 4.2 is valid for gentle hills only. Steeper hills and mountains lead to non-linear features such as flow separation and other features named in Sect. 4.1 which are not adequately covered by the equations given in Sect. 4.2. Non-linear flow features can no longer be derived from analytical relations but require the operation of numerical flow models. Some non-linearity effects become already visible in the examples for the flow over an escarpment in Sect. 4.3. Therefore, wind assessment in rougher terrain where linearity is no longer assured has to be done by site-specific numerical model simulations. This Chapter was designed to point to the main flow features which influence the vertical wind profile over hills and gentle mountains.

References

Allnoch, N.: Windkraftnutzung im nordwestdeutschen Binnenland: Ein System zur Standortbewertung für Windkraftanlagen. Geographische Kommission für Westfalen, Münster, ARDEY-Verlag, 160 pp. (1992).

Anderson P.S., Ladkin R.S., Renfrew I.A.: An Autonomous Doppler Sodar Wind Profiling System. J. Atmos. Oceanic Technol. **22**, 1309–1325 (2005).

Astley, R.J.: A Finite Element Frozen Vorticity Solution for Two-Dimensional Wind Flow over Hills. 6th Australasian Conf. on Hydraulics and Fluid Mechanics, Adelaide, Australia, 443–446 (1977).

Atkinson B.W.: Meso-scale Atmospheric Circulations. Academic Press, London etc., 495 pp. (1981).

Bowen, A.J., D. Lindley,: A Wind-Tunnel Investigation of the Wind Speed and Turbulence Characteristics Close to the Ground over Various Escarpment Shapes. Bound.-Layer Meteorol. **12**, 259–271 (1977).

Bowen, A.J.: Full Scale Measurements of the Atmospheric Turbulence over Two Escarpments. In: J.E. Cermak (ed.), Wind Engineering: Proc. 5th Internat. Conf., Fort Collins, Pergamon, 161–172 (1979).

Bradley, E. F.: The Influence of Thermal Stability and Angle of Incidence on the Acceleration of Wind up a Slope. J. Wind Eng. Indust. Aerodynam. **15**, 231–242 (1983).

Caccia, J.-L., Guénard, V., Benech, B., Campistron, B., Drobinski, P.: Vertical velocity and turbulence aspects during Mistral events as observed by UHF wind profilers. Ann. Geophysicae 22, 3927–3936 (2004).

Defant, F.: Zur Theorie der Hangwinde, nebst Bemerkungen zur Theorie der Berg- und Talwinde. Arch. Meteorol. Geophys. Bioklimatol. A 1, 421–450 (1949).

Emeis, S., H.P. Frank, F. Fiedler: Modification of air flow over an escarpment—Results from the Hjardemal experiment. Bound.-Lay. Meteorol. **74**, 131–161. (1995).

Emeis, S.: Vertical variation of frequency distributions of wind speed in and above the surface layer observed by sodar. Meteorol. Z. **10**, 141–149 (2001).

Founda, D., M. Tombrou, D.P. Lalas, D.N. Asimakopoulos: Some measurements of turbulence characteristics over complex terrain. Bound.-Lay. Meteorol. **83**, 221–245 (1997).

Frank, H., K. Heldt, S. Emeis, F. Fiedler: Flow over an Embankment: Speed-Up and Pressure Perturbation. Bound.-Lay. Meteorol. **63**, 163–182 (1993).

Heimann, D., De Franceschi, M., Emeis, S., Lercher, P., Seibert, P. (Eds): Air pollution, traffic noise and related health effects in the Alpine space—a guide for authorities and consulters. ALPNAP comprehensive report. Università degli Studi di Trento, Trento, 335 pp. (2007).

Hoff, A.M.: Ein analytisches Verfahren zur Bestimmung der mittleren horizontalen Windgeschwindigkeiten über zweidimensionalen Hügeln. Ber. Inst. Meteorol. Klimatol. Univ. Hannover, **28**, 68 pp. (1987).

Jackson, P.S., J.C.R. Hunt: Turbulent wind flow over a low hill. Quart. J. Roy. Meteorol. Soc. **101**, 929–955 (1975).

Jensen, N.O.: A Note on Wind Generator Interaction. Risø-M-2411, Risø Natl. Lab., Roskilde (DK), 16 pp. (1983) (Available from http://www.risoe.dk/rispubl/VEA/veapdf/ris-m-2411.pdf).

Jensen, N.O., Petersen, E.L., Troen, I.: Extrapolation of Mean Wind Statistics with Special Regard to Wind Energy Applications, Report WCP-86, World Meteorol. Organization, Geneva, 85 pp. (1984).

Justus, C.G., W.R. Hargraves, A. Mikhail, D. Graber: Methods for Estimating Wind Speed Frequency Distributions. J. Appl. Meteor. **17**, 350–353 (1978).

Lugauer, M,, Winkler, P.: Thermal circulation in South Bavaria—climatology and synoptic aspects. Meteorol. Z. **14**, 15–30 (2005).

Mason, P. J.: Flow over the Summit of an Isolated Hill, Bound.-Lay. Meteorol. **37**, 385–405 (1986).

Panofsky, H.A., D. Larko, R. Lipschutz, G. Stone, E.F. Bradley, A.J. Bowen und J. Højstrup: Spectra of velocity components over complex terrain. Quart. J. Roy. Meteorol. Soc. **108**, 215–230 (1982).

Petersen, E.L., N.G. Mortensen, L. Landberg, J. Højstrup, H.P. Frank: Wind Power Meteorology. Part II: Siting and Models. Wind Energy, **1**, 55–72 (1998b).

Renfrew, I.A., Anderson, P.S.: Profiles of katabatic flow in summer and winter over Coats Land, Antarctica. Quart. J. Roy. Meteor. Soc. **132**, 779–802 (2006).

Smith, R.B.: The influence of mountains on the atmosphere. In: Landsberg HE, Saltzman B (Eds) Adv. Geophys. 21, 87–230 (1978).

Steinacker, R.: Area-height distribution of a valley and its relation to the valley wind. Contr. Atmos. Phys. **57,** 64–71 (1984).

Sykes, R.I.: An Asymptotic Theory of Incompressible Turbulent Boundary Layer Flow over a Small-Lump. J. Fluid Mech. **101**, 647–670 (1980).

Taylor, P.A.: Numerical studies of neutrally stratified planetary boundary layer flow over gentle topography, I: Two-dimensional cases. Bound.-Lay. Meteorol., **12**, 37–60 (1977).

Taylor, P.A., Mason, P.J., Bradley, E.F.: Boundary-Layer Flow over Low Hills. Bound.-Lay. Meteorol. **39**, 107–132 (1987).

Troen, I., E.L. Petersen: European Wind Atlas. Risø National Laboratory, Roskilde, Denmark. 656 pp. (1989).

Vergeiner, I.: An energetic theory of slope winds. Meteorol. Atmos. Phys. **19**, 189–191 (1982).

Vergeiner, I., Dreiseitl, E.: Valley winds and slope winds—observations and elementary thoughts. Meteorol. Atmos. Phys. **36**, 264–286 (1987).

Zenman, O., N.O. Jensen: Modification of Turbulence Characteristics in Flow over Hills. Quart. J. Roy. Meteorol. Soc. **113**, 55–80 (1987).

Chapter 5
Offshore Winds

This Chapter deals with the marine atmospheric boundary (MABL). The special features of wind and turbulence profiles over the sea are very important, since an increasingly larger fraction of wind energy will be generated at offshore wind parks in the future. Although the sea surface is perfectly flat, these wind features partly differ from profiles over homogeneous land presented in Chap. 3. Unless otherwise stated, the examples for the state of the MABL presented in this Chapter are based on Türk (2008). Türk (2008)'s analysis was based on data from the 100 m tower FINO1 in the German Bight. This offshore tower, which is about 45 km away from the German coast, provides wind information from cup anemometers in heights between 30 and 100 m with a vertical resolution of 10 m. Sonics data are available at 40, 60 and 80 m from this tower. So, some of the presented features may be specific for the German Bight at the site of FINO1. Nevertheless, they can serve as an indication for typical behaviour in the MABL in contrast to an onshore boundary layer. There are more measurement towers near offshore wind parks, e.g., the 62 and 70 m masts at Horns Rev off the Danish west coast or the 116 m mast "NoordzeeWind" off the Dutch coast near Egmond aan Zee. In Germany there are the towers FINO2 in the Baltic and FINO3 in the German Bight off the island of Sylt as well. These latter two towers are quite similar to the tower FINO1.

Section 5.1 explains the special features of the sea surface. Section 5.2 then presents mean vertical profiles before Sect. 5.3 deals with extreme wind speeds and Sect. 5.4 with turbulence parameters in the MABL. Weibull parameters characterizing the marine boundary layer are discussed in Sect. 5.5. In coastal areas, which are the subject of Sect. 5.6, internal boundary layers (see Sect. 3.5) can form which exhibit marine boundary characteristics in the internal boundary layer and onshore boundary layer characteristics in the layer above. Especially for stable thermal stratification when warmer air is advected over colder water, such internal boundary layers can persist for long distances of several tens of kilometres.

S. Emeis, *Wind Energy Meteorology*, Green Energy and Technology,
DOI: 10.1007/978-3-642-30523-8_5, © Springer-Verlag Berlin Heidelberg 2013

5.1 Characteristics of Marine Boundary Layers

First of all, the sea surface is much smoother than the land surface. This leads to higher wind speeds at a given height above the surface, to smaller turbulence intensities and to shallower surface layer depths. Thus, offshore wind turbines usually experience less wind shear over the rotor area. But sea surface roughness is wind speed-dependent due to the formation of waves. Diurnal cycles of temperature and atmospheric stability are nearly absent due to the large heat storage capacity of water. The infinite moisture source at the sea surface tends to bias static stability of the MABL towards unstable stratifications. Figure 5.1 gives the principle features of the vertical structure of the MABL. Adjacent to the sea surface we find the wave sublayer within which the direct influence of single waves through pressure forces is dominant. This sublayer is roughly five wave amplitudes deep. Above the wave sublayer we find the constant flux or Prandtl layer which is often much shallower than the respective layer over land (see Fig. 3.1). This depth can be in the order of just 10 m for stable stratification and in low to moderate winds. The upper 90 % of the MABL are covered by the Ekman layer within which the wind slightly turns and reaches the geostrophic wind at its top. Like the constant flux layer the entire MABL is usually much shallower than the ABL over land.

5.1.1 Sea Surface Roughness and Drag Coefficient

The typical roughness length of the sea surface for moderate wind speeds is in the order of a tenth of a millimetre to a millimetre (see Fig. 5.2 left). In contrast to land surfaces the roughness of the sea surface is not constant but varies over several decades depending strongly on the wind speed, because of the evolving wave size, height and shape. Consequently, the surface roughness length, z_0 increases with increasing wind speed. Waves are generated mainly by frictional forces exerted by the wind on the ocean surface, thereby transporting momentum from the atmosphere downwards into the water column (Bye and Wolff 2008). This transport is downward as long as the waves are still young and wind-driven, i.e. if the wind speed is faster than the phase speed of the waves. For old waves or swell, no clear relation with the wind speed can be expected (Oost et al. 2002; Sjöblom and Smedman 2003). Furthermore, this downward transport depends also on the thermal state of the MABL, because this state influences the ability of the atmosphere to replenish the momentum loss at its lower boundary with momentum from higher atmospheric layers. For unstable stratification (air colder than the sea), this downward transport is larger and the waves are expected to be higher than for stable stratification. This presumption has initially been proven by the analysis of North Atlantic weather ship data by Roll (1952).

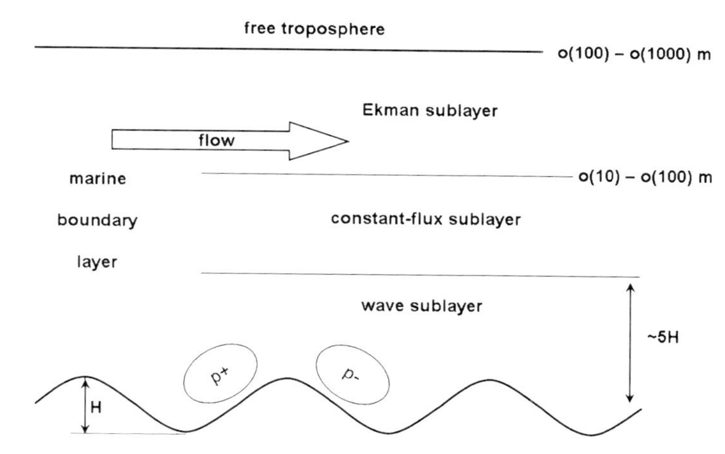

Fig. 5.1 Vertical structure of the marine boundary layer over a wavy sea surface. p^+ and p^- indicate positive and negative pressure perturbations close to the waves

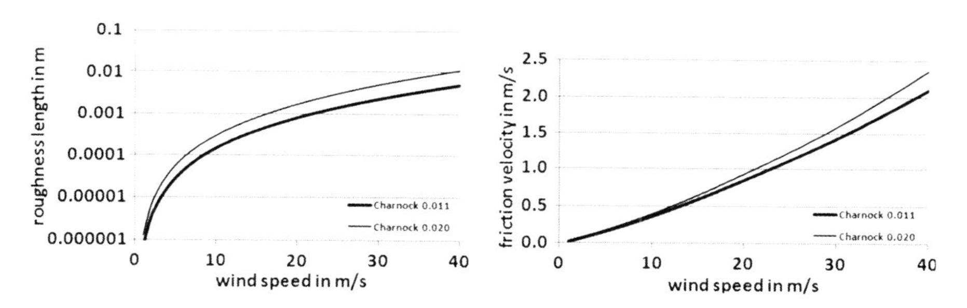

Fig. 5.2 Roughness length of the sea surface in m (*left*) and friction velocity in m/s (*right*) using Charnock's relation (5.1) and the neutral logarithmic wind profile (3.6) for two different values of the Charnock parameter (*bold line*: 0.011, *thin line*: 0.020)

Many studies on the wind-driven roughness of the sea surface already exist. Charnock (1955) presented a relation between roughness length, z_0 and friction velocity, u_* based on a small dataset collected under near-coastal conditions at a measurement height of eight metres:

$$z_0 = \frac{\alpha u_*^2}{g} = \frac{\alpha \kappa^2 u(z)^2}{g\left(\ln\frac{z}{z_0} - \Psi(\frac{z}{L_*})\right)^2} \tag{5.1}$$

where z_0 is the surface roughness length, u_* the friction velocity and g the acceleration of gravity. The term behind the second equal sign in (5.1) has been derived by using the formula for the diabatic wind profile (3.16). This latter relation has to be solved iteratively. The empirical constant, α is called today the Charnock parameter. For the open ocean Smith (1980) suggests $\alpha = 0.011$ while at shallow or near-coastal sites α is a little larger with values about 0.016–0.02 (Garratt 1977; Wu 1980). Garratt (1977) summarized sea surface drag coefficients from 17 experiments and supported Charnock's relation. Using a friction velocity of 0.33 m/s and $\alpha = 0.018$ gives $z_0 = 0.00018$ m.

The determination of the sea surface drag coefficient, C_D is another way of looking at sea surface roughness. The drag coefficient for neutral atmospheric stability and 10 m height is defined as:

$$C_{DN10} = \frac{u_*^2}{\overline{u}_{10}^2} \tag{5.2}$$

where u_* is the friction velocity defined in (3.16) and $\overline{u}_{10}$ is the 10 m wind speed defined in (3.1). Despite conflicting evidence in the past (Garratt 1977), it is now accepted that the drag coefficient in the MABL is an increasing function of the wind speed (Sullivan and McWilliams 2010) for moderate wind speeds (see Fig. 5.4). This becomes obvious when inserting the logarithmic wind profile (3.6) for the denominator of (5.2) using (5.1) for the determination of the roughness length:

$$C_{DN10} = \left(\frac{\kappa}{\ln \frac{gz}{\alpha u_*^2}} \right)^2 \tag{5.3}$$

where $z = 10$ m. At higher wind speeds, however, most data sets suggest that the drag coefficient tends toward a constant value (Anderson 1993; Donelan et al. 2004; Black et al. 2007). A few, e.g., the HEXOS data (Janssen 1997, triangles in the left frames of Figs. 5.3 and 5.4) do not show this levelling off. The exact equation that describes the relationship between the drag coefficient and wind speed is dependent on the author (Geernaert 1990). Although an universal consensus does not exist, the most widely cited relationships are possibly those proposed by Smith (1980):

$$C_{DN10} = 0.00061 + 0.000063\overline{u}_{10} \tag{5.4}$$

said to be valid for a wind speed range between 6 and 22 m/s, the one proposed by Large and Pond (1981):

$$C_{DN10} = \begin{matrix} 0.00114 & \text{for } 4\text{ m/s} < u_{10} \leq 10\text{ m/s} \\ 0.00049 + 0.000065u_{10} & \text{for } 10\text{ m/s} < u_{10} < 26\text{ m/s} \end{matrix} \tag{5.5}$$

and the one by Yelland et al. (1998):

$$C_{DN10} = 0.00050 + 0.000071\overline{u}_{10} \tag{5.6}$$

said to be valid for a wind speed range between 6 and 26 m/s. Similar wind speed dependencies come from the Coupled Ocean–Atmosphere Response Experiment (COARE) algorithm by Fairall et al. (1996, 2003).

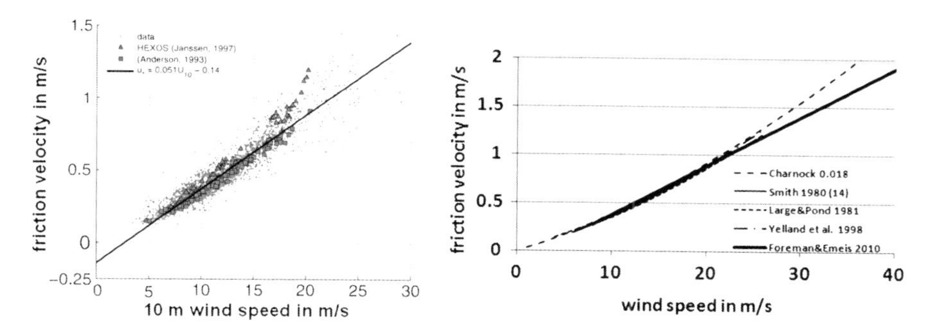

Fig. 5.3 Friction velocity in the marine surface layer, u_* plotted against 10 m wind speed, u_{10}. *Left*: from literature data listed in Table 1 of Foreman and Emeis (2010). For $u_{10} > 8$ m/s and $u_* > 0.27$ m/s a *straight line* [see Eq. (5.8)] is fitted to data in this range. The HEXOS results as reported by Janssen (1997) are shown by *triangles*; the measurements of Anderson (1993) are indicated by *squares* (from Foreman and Emeis 2010). Right: functional dependencies of friction velocity on wind speed: *bold line*: Eq. (5.8), *dashed line*: Eq. (5.3) using $\alpha = 0.018$, *thin line*: Eq. (5.4), *dotted line*: Eq. (5.5), *dash-dotted line*: Eq. (5.6)

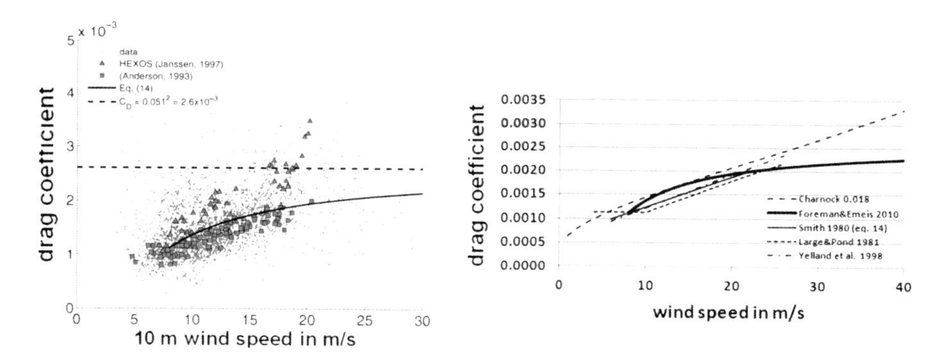

Fig. 5.4 Drag coefficient of the sea surface, C_D plotted against 10 m-wind speed, u_{10}. *Left*: from literature data listed in Table 1 of Foreman and Emeis (2010). For $u_{10} > 8$ m/s and $u* > 0.27$ m/s a curve [see Eq. (5.9)] is fitted to data in this range. The HEXOS results as reported by Janssen (1997) are shown by *triangles*; the measurements of Anderson (1993) are indicated by *squares* (from Foreman and Emeis 2010). *Right*: functional dependencies of drag coefficient on wind speed: *bold line*: Eq. (5.9), *dashed line*: Eq. (5.3) using $\alpha = 0.018$, *thin line*: Eq. (5.4), *dotted line*: Eq. (5.5), *dash-dotted line*: Eq. (5.6)

Differences in measured drag coefficients between independent studies are most probably a function of the state of the sea (Donelan 1990) such as the wave steepness or slope (e.g., Hsu 1974) and wave age (e.g., Maat et al. 1991). For example, the drag coefficient is thought to increase with younger waves (i.e. decreasing wave age) (Smith et al. 1992). The precise dependence of the drag coefficient on one or more of these tools is an ongoing area of research in air-sea interaction (Sullivan and McWilliams 2010).

Nevertheless, a wind speed-dependent drag coefficient is not desirable, because usually a drag coefficient for a fully turbulent flow should be only object-dependent and independent of the flow speed above this object. Only then flows at different speeds are similar to each other. A wind speed-dependent drag coefficient indicates that the state of flow is changing with wind speed. Most probably, the flow over the very smooth sea surface is not fully turbulent for 10 m wind speeds of less than about 8 m/s. The drag coefficient should be a constant above this wind speed.

Therefore, a new functional form of the neutral drag coefficient for moderate to high wind speeds in the MABL for a range of field measurements as reported in the literature has been proposed by Foreman and Emeis (2010). This new form is found to describe a wide variety of measurements recorded in the open ocean, coast, fetch-limited seas, and lakes, with almost one and the same set of parameters. It is the result of a reanalysis of the definition of the drag coefficient in the marine boundary layer, which finds that a constant is missing from the traditional definition of the drag coefficient. The constant arises because the neutral friction velocity over water surfaces is not directly proportional to the 10 m wind speed, a consequence of the transition to rough flow at low wind speeds below about 8 m/s. Within the rough flow regime, the neutral friction velocity is linearly dependent on the 10 m wind speed; consequently, within this rough regime, the newly defined drag coefficient is not a function of the wind speed. The magnitude of the newly defined neutral drag coefficient represents an upper limit to the magnitude of the traditional definition.

In order to derive this new wind speed independent drag coefficient, Foreman and Emeis (2010) start with an analysis of the relation between the friction velocity and wind speed. Solving (5.2) for the friction velocity gives:

$$u_* = \sqrt{C_{DN10}}\overline{u}_{10} \tag{5.7}$$

(5.7) does not depict the reality, especially not for higher wind speeds. A better relation is (straight line in Fig. 5.3 left and bold line in Fig. 5.3 right):

$$u_* = \sqrt{C_{mN10}}\overline{u}_{10} + b \tag{5.8}$$

with $C_{mN10} = 0.0026$ and $b = -0.14$ m/s. The straight line described by (5.8) does not meet the origin, thus it is valid only for the fully turbulent regime above 8 m/s wind speed. Inserting (5.7) into (5.1) yields:

$$C_{DN10} = \frac{\left(\sqrt{C_{mN10}}\overline{u}_{10} + b\right)^2}{\overline{u}_{10}^2} \tag{5.9}$$

This new relation (5.9) is depicted as bold curves in Fig. 5.4. For high wind speeds, the classical drag coefficient C_{DN10} from (5.9) converges against C_{mN10}. $C_{mN10} = 0.0026$ is shown as dashed horizontal line in Fig. 5.4 left.

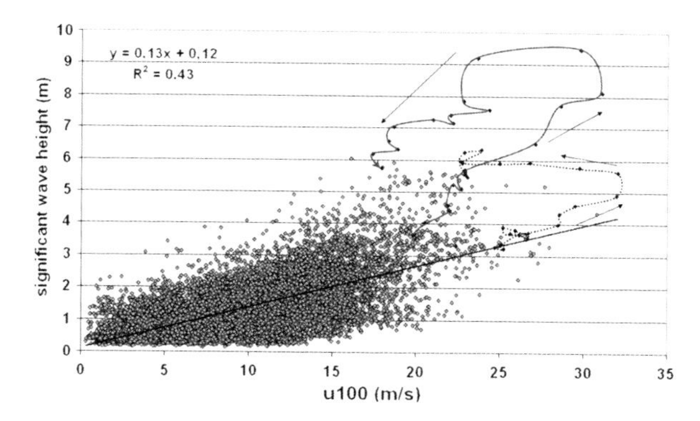

Fig. 5.5 Relation between significant wave height, H_s in m and 100 m wind speed in m/s in the German Bight from three years of hourly data (2004–2006) at FINO1. The two curves indicate the temporal development of H_s for storm "Britta" on November 1, 2006 (*full curve*) and for storm "Erwin" on January 8, 2005 (*dotted curve*). *Arrows* indicate the direction of this development

5.1.2 Fetch and Stability Dependent Wave Formation

The preceding subchapter has shown that the development of waves is decisive for the sea surface roughness. The calculation of oceanic wave heights from local wind speeds has a long and well-established history in oceanic and atmospheric sciences (see, e.g., Sverdrup and Munk 1947; Neumann 1953). Although the local wind speed and the local structure of the MABL are supposed to have an important influence, there are other factors determining the wave height. Wave heights additionally depend on the length of the fetch and on the duration of high wind speeds over these fetches. Further on, the wave height also depends on non-atmospheric conditions like the, e.g., simulations with the wave ocean model (WAM; Hersbach and Janssen 1999) have shown for infinite duration and deep water that for a wind speed of 30 m/s, the significant wave height increases from about 10 m for 50 km fetch to nearly 12 m for 100 km fetch and more than 15 m for 400 km fetch.

A closer analysis of development of wind-driven wave heights in the German Bight can be found in Emeis and Türk (2009). In this study wind speed, friction velocity and significant wave height data from the FINO1 platform in the southern German Bight 45 km off the coast for the years 2004–2006 have been evaluated and related to each other. The wave height is usually expressed in terms of a significant wave height, H_s. H_s is often defined as the average height (trough to crest) of that third of waves out of all waves which have the largest wave heights. The maximum wave height is 1.6–1.7 times the significant wave height (Kumar et al. 1999). Toba (1978) has given a relation between the friction velocity, u_*, and

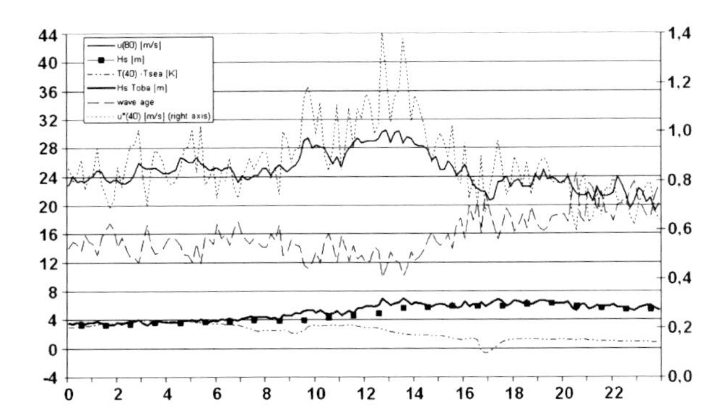

Fig. 5.6 Wind speed at 80 m (*full line*) and friction velocity, u_*, at 40 m (*dotted*) above mean sea level together with the wave age (*dashed line*), the air–sea temperature difference (*dashed-double dotted line*), the measured hourly wave height (*black squares*) and the calculated wave height from Eq. (5.10) (*thick full line*) at FINO1 in the German Bight during the violent storm "Erwin" on January 8, 2005 (x-axis gives time in hours). All data except measured wave heights are 10 min mean data. u_* refers to the *right*-hand axis, all other variables to the *left*-hand axis

the significant wave height, H_s, for growing waves independent of the fetch if the wave period T is known:

$$H_s = 0.062\sqrt{u_* g T^3} \tag{5.10}$$

with g the acceleration of gravity. Maat et al. (1991) give a somewhat smaller value for the constant in Eq. (5.10), namely 0.051.

Figure 5.5 shows a broad scatter of wave height data with wind speed. The two curves in Fig. 5.5 show the "trajectories" for the storms "Britta" and "Erwin" in the wave height–velocity phase space. These trajectories demonstrate that a larger part of the scatter in Fig. 5.5 happens with the evolution of the wave height–velocity relation during the passage of single low-pressure systems. The curves are plotted through consecutive hourly data points. The full curve for the All Saints Day storm "Britta" is 21 h long from October 31, 2006, 1700 h to November 1, 1400 h. The second curve for the gale force storm "Erwin" covers a time span of 23 h from January 7, 2005, 2300 h to January 8, 2200 h. This is analysed in more detail in Figs. 5.6 and 5.7. During "Britta" (Fig. 5.7), the peak H_s is reached about 1 h after the peak wind speed of 31 m/s; during "Erwin" (Fig. 5.6), it is reached about 5 h after the main wind speed peak of 32 m/s. The two curves for "Britta" and "Erwin" differ considerably. The curve for "Britta" lies at much higher wave heights than the curve for "Erwin", although the peak wind speeds are quite similar. Looking at the wind direction and the air-sea temperature difference, it turns out that the major difference in atmospheric conditions is that during "Britta", cold air advection was prevailing with northerly winds, and during "Erwin", warm air advection with westerly winds.

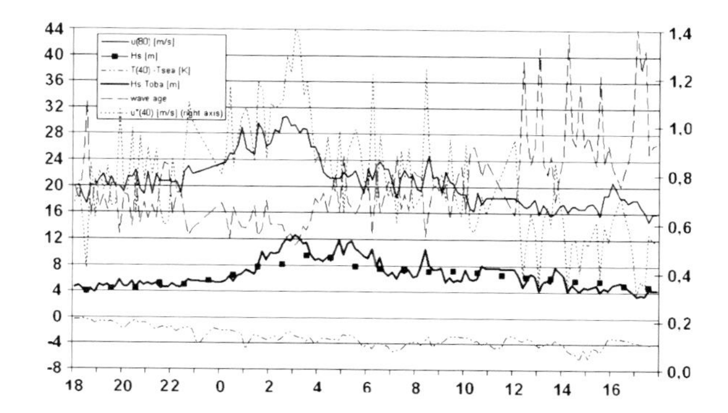

Fig. 5.7 As Fig. 5.6, but for violent storm "Britta" on October 31/November 1, 2006. Between 11 p.m. and midnight and between 11 a.m. and noon, some data are missing

Figures 5.6 (Erwin) and 5.7 (Britta) analyse the relation between the local state of the atmosphere and the wave height in these two storm events in some more detail also using some of the 10 Hz data from the sonic anemometers at 40 and 80 m. The atmospheric conditions are described by the wind speed, u, at 80 m and the friction velocity, u_*, at 40 m height. In addition, the temperature difference between the air temperature at 40 m height and the sea surface temperature, the measured hourly values of the significant wave height, H_s, the calculated significant wave heights using Eq. 5.10 and the wave age (i.e. the phase speed of the waves over the friction velocity, see also next subchapter) are given. Figure 5.6 shows a slightly stably stratified boundary layer due to warm air advection over colder water (see temperature difference). The mean ratio, z/L_*, is only about +0.03 (about 0.06 in the first half of the displayed period and nearly 0.0 in the second half, not shown in the figure). The vertical wind shear between 40 and 80 m is about 3 m/s and decreases from 1 to 2 m/s after the occurrence of the peak wind velocity. The peak wind speed is accompanied by a maximum in the friction velocity (1.4 m/s) and a minimum in the wave age (about 11). The greatest increase in the wave height coincides with the highest values of the friction velocity. The peak wave height is observed 5 h after the peak wind speed at a wave age of about 18. For the whole period shown in Fig. 5.6, the wave age remains below 24, i.e. we have a wind-driven sea all the time. Figure 5.7 shows an unstably stratified boundary layer during cold air advection of warmer water. The mean ratio, z/L_*, is only about −0.06 (between midnight and 11 a.m.). Negative peak values of z/L* of up to −0.79 occur between 2 and 4 p.m. (not shown in the figure). There is nearly no vertical wind shear in the layer between 40 and 80 m. Also, the friction velocity is nearly constant with height. Thus, this layer seems to be a constant flux layer (Fig. 5.1). Again, the largest increase in wave height coincides with the phase of the highest friction velocity (again 1.4 m/s). The peak wave height is observed about 1 h after the peak wind speed at a wave age of about 11–12. Like in Fig. 5.6, the wave age shows a minimum associated with the maximum in the friction velocity, u_*, but in contrast to the situation during storm

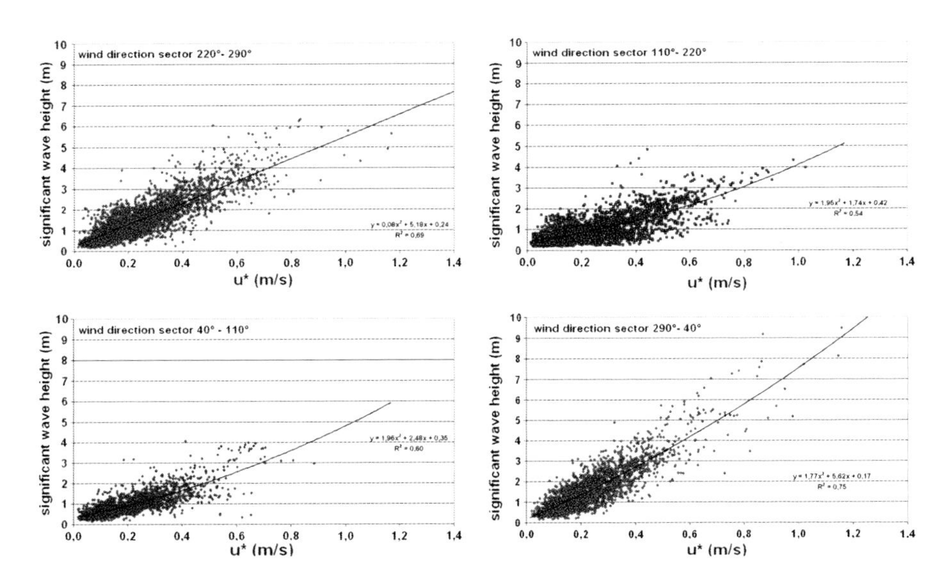

Fig. 5.8 Significant wave height in m versus friction velocity, u_*, at 40 m (hourly data) for westerly (*upper left*), southerly (*upper right*), easterly (*lower left*), and northerly (*lower right*) winds at FINO1 in the German Bight

"Erwin", the wave age is changing from young to older waves about 10 h after the passage of the peak wind speed at FINO1. In general, we find in both cases an anticorrelation between friction velocity and wave age.

Because the values for the stability parameter z/L_* for the two cases are so close together, the values for the friction velocity are very similar in both cases. The calculated wave heights from Eq. 5.10 in Figs. 5.6 and 5.7 have been produced by choosing 0.056 as a value for the constant in Eq. 5.10 because this value gives the best fit. This value turns out to be between the ones proposed by Maat et al. (1991) and by Toba (1978). It becomes obvious that the calculated wave height is above the measured one as long as the wave height is increasing due to the shear stress exerted by the atmosphere on the sea surface. The periods of overestimation from (5.10) coincide with wave ages close to 12 or even lower. After having reached the peak wave height, the calculated wave height is slightly lower than the measured one in Fig. 5.7. This systematic deviation after the peak wave height—which becomes especially notable in Fig. 5.7 for wave ages over 24 (equilibrium to old waves)—is meaningful because Toba's relation has been derived for growing waves only. On the other hand, the overall comparison between measured and calculated wave heights turns out quite well and therefore provides an independent confirmation of the values for the friction velocity, u_*, determined from the sonic anemometer measurements.

It is not meaningful to derive a relation between wave height and wind speed from the data plotted in Fig. 5.5 due to the large scatter. Therefore, Fig. 5.8 presents the data separately for four different wind sectors (see Table 5.1 for exact

Table 5.1 Significant wave heights, H_s for four wind direction sectors and stratification as function of selected wind speed values at 100 m height from respective regression curves to those shown in Fig. 5.8 (taken from Emeis and Türk 2009)

Wind speed in m/s	25	30	35	40	Explained variance in %
Sector/stratification	Significant wave height in m				w.r.t. wind speed
North (290°–40°, usually unstable)	6.7	9.2	12.1	15.5	69.5
East (40°–110°), unstable	5.2	7.0	9.2	11.8	75.2
South (110°–220°), unstable	4.2	6.0	8.1	10.6	61.3
South (110°–220°), stable	3.4	4.8	6.5	8.5	53.0
West (220°–290°, usually stable)	4.1	5.3	6.6	8.1	56.6
East (40°–110°), stable	1.6	1.6	1.7	1.6	29.9

The entries in this table are ordered with decreasing H_s for $u = 40$ m/s

definition of these sectors). The relatively low wave heights from the eastern and especially from the southern wind direction sector have to be attributed to the small fetches and therefore limited durations for which wind and waves can interact in these two sectors. The minimum distance to the coast in the southern sector is only about 50 km. Fetches are much longer for the western and the northern sectors and reach or even exceed the spatial scales of atmospheric depressions. Therefore, durations of 12–24 h can be assumed for these two sectors. As waves are higher in the northern sector than in the western sector for the same observed friction velocity, the waves from the northern sector must be older than the waves from the western sector. Because wave periods were not easily available the regression between the wave heights and the friction velocity has been approximated by a quadratic expression in the four frames of Fig. 5.8.

The wave heights in the different wind direction sectors have also been correlated to the wind speed in 100 m height. The results are given in Table 5.1. Two features become obvious: (1) The explained variance is somewhat lower when the wave heights are correlated with the wind speed than with friction velocity, and (2) for two sectors (east and south), the thermal stratification of the MABL becomes important. Therefore, different regressions for stable and unstable stratification have been listed in Table 1. In these two sectors, the land is still so close to the measurement site that air considerably warmer (stable stratification) or colder (unstable) can reach the FINO1 platform. The vertical mixing is not sufficient to remove this stratification on the way from the coast to the platform. For the east wind sector, this difference between stable and unstable stratification is depicted in Fig. 5.9. From this sectoral analysis, the highest wave heights at FINO1 in the German Bight have to be expected from the northern sector.

The missing stratification dependence for westerly and northerly winds in Table 5.1 indicates that stronger winds from the western and the northern sector at FINO1 in the German Bight are linked to a limited range of possible thermal stratifications. Northerly gale force winds at this site occur mainly during cold air outbreaks on the rear side of cyclones moving east over Northern Europe,

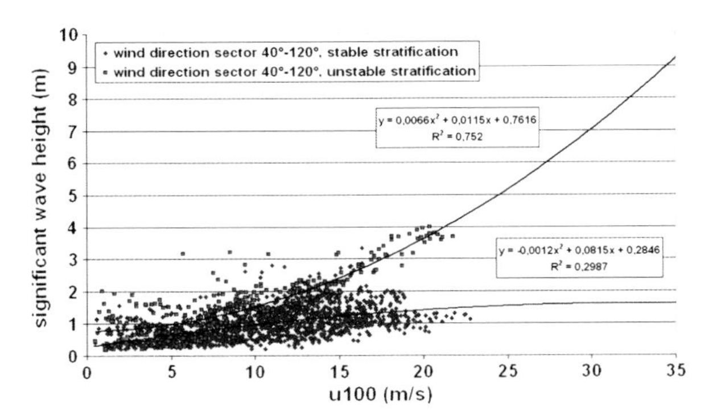

Fig. 5.9 Similar to lower left frame of Fig. 5.8, but plotted versus wind speed. Distinction has been made between stable stratification (*small diamonds*) and stable stratification (*larger squares*)

whereas westerly gale force winds usually occur within a warm sector of cyclones moving towards Northeast or East. This finding is supported by looking at the air-sea temperature difference for the two cases displayed in Figs. 5.6 and 5.7. During the All Saints Day storm "Britta" on November 1, 2006, the winds came from the northern sector (cf. Fig. 5.8 lower right) and the air temperature was several degrees lower than the sea surface temperature. Therefore, this was a case with unstable stratification. During the passage of the cyclone "Erwin" on January 8, 2005, the air temperature was somewhat higher than the sea surface temperature, indicating a slightly stable stratification. Both storms brought extreme wind speeds, but for the flow pattern and the thermal stratification, they were typical for higher winds from these sectors. This is the reason why we do not find notable differences between stable and unstable situations in these two sectors in the way we had found it in the other two sectors. Thus, the northern sector can be seen as a selection of weather situations with usually unstable stratification and the western sector as a selection with usually stable stratification, at least in cases with stronger winds. This stratification difference between these two sectors explains why the wave heights in the western sector are much lower than in the northern sector, although fetch and duration are large in both sectors.

5.1.3 Extreme Wave Heights

Emeis and Türk (2009) also estimated the possible extreme wave heights for the four sectors in Fig. 5.8 and Table 5.1 using the techniques described in Appendix A.3 after Eq. (A.32). In Fig. 5.10 we plotted the cumulative frequencies of all wave heights (in 1 m bins) in the different sectors (keeping the differentiation for thermal stability in the southern and eastern sector). The 50 year threshold in this plot refers to about 2,000 values a year which correspond to the number of data per year in the most populated wind direction sector [for $N = 2{,}000$ the 50 year threshold $y = -\ln(-\ln(1 - 1/(50 \times N)))$ is about 11.5]. Although the most frequent wind direction is from southwest, the most populated wind sector is the

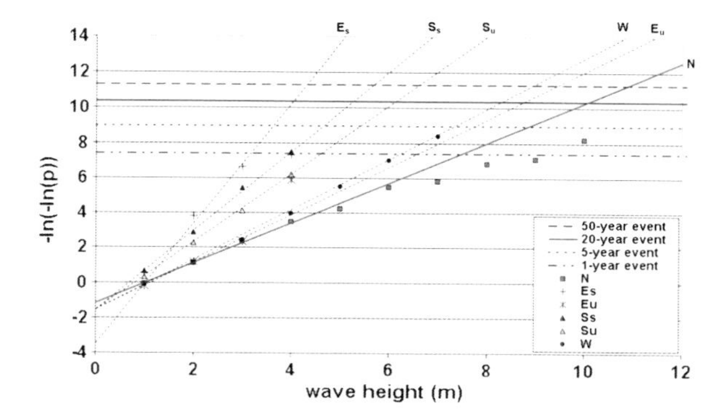

Fig. 5.10 Estimation of extreme wave heights for different wind sectors at FINO1. Capital letters denote wind direction (see Table 5.1), subscripts "s" and "u" denote stable and unstable thermal stratification. *Slanted lines* are approximations to the data points giving more weight to lower wave heights. The y-axis refers to the data from the northern sector (*full slanted line*); data from the other sectors (*dotted slanted lines*) have been shifted vertically accordingly to match the plotted 1–50 year thresholds

northern sector because with 110°, it is much wider than the western sector with only 70°. Curves from the other sectors, which are based on fewer values, have 50 year thresholds at y values lower than 11.5. These curves have therefore been shifted vertically accordingly so that their thresholds match the horizontal lines in Fig. 5.10 which indicate the thresholds for the northern sector. The highest 50 year extreme significant wave heights which have to be expected will probably come from the northern, eastern (under unstable conditions, i.e. cold air advection) and from the western sector with 9–11 m. Extreme significant wave heights from the southern sector and the eastern sector under stable conditions will only be between 4 and 7 m. The uncertainty of this extreme value estimation can be assessed from the plots. The better the data fit to a Fisher–Tippett type 1 distribution, the more the data points should arrange themselves along a straight line in the plot.

The uncertainty of the 50 year extreme value could be estimated from the spread of the crossing points between possible straight lines through the data and the 50 year line ($y = 11.5$). From this criterion, the uncertainty for the western sector is the smallest (a few percent only). For the other sectors (except the northern one), it may be up to about 10 %. For the northern sector, the reliability is also good if values up to 6 m wave height are used, which has actually been done in Fig. 5.10. The data points for the northern sector in Fig. 5.10 for wave heights above 6 m all come from the storm "Britta". The deviation of these data points to the right from the straight line indicates that "Britta" must have been a rather rare event. Taking the highest value (10 m) and going straight upright, we hit the straight regression line for the Gumbel distribution at about the probability for the 20 year event. Thus, following the analysis given here, "Britta" has been a 20 year event, whereas "Erwin" was not unusual and can be expected every 1–3 years. This extreme value estimation technique is not devalued due to the fact that about

20 consecutive values from the storm "Britta" have entered the analysis. This duration of a storm event has been considered as a typical duration of an atmospheric depression. As there are a lot of other (weaker) storms with comparable durations also included in the analysis, the weight of "Britta" in this analysis can be considered as fair.

5.1.4 Wave Age

The wave age is an important parameter which governs vertical profiles of wind and turbulence in the constant flux layer of the MABL. The wave age describes the type of interaction between the wind field and the waves. A distinction is made between young and old waves. This distinction is necessary, because of the delayed response of the wave field to the wind field and the hysteresis effects which come with this delay. Young waves are wind-driven waves where the wind speed is larger than the phase speed of the waves. This situation resembles the usual situation over rough land surfaces and we expect principally a validity of Monin–Obukhov similarity and the features described in Chap. 3. Old waves are waves which still exist after the wind force has already decreased again. Old waves are often called swell when they come in from far away. These waves can be faster than the near-surface wind, which essentially means that the waves drive the near-surface wind and that upward turbulent momentum flux can occur in the surface layer of the MABL. This behaviour can no longer be described by Monin–Obukhov similarity, because this similarity approach assumes that the surface is a momentum sink. The influence of wave age will become visible in some of the results in the following subchapters.

Typically, wave age, c is defined as the ratio between the phase speed of the waves, c_{ph} and the friction velocity in the atmospheric surface layer:

$$c = \frac{c_{ph}}{u_*} \tag{5.11}$$

The limit between young and old wave is roughly at 28, because the friction velocity is in the order of 1/28 of the wind speed. Typical values for the wave age range from 5 at high wind speeds to several hundred at very low wind speeds. Figure 5.11 shows the relative frequency distribution for 2005 observed at the FINO1 platform. The most frequent wave age is between 25 and 30, i.e. just around the transit threshold between young and old waves. The average wave age is much higher at 55.3 due to fewer but very large wave ages. Figure 5.12 displays the relation between wind speed and wave age. As said above, young waves occur at high wind speeds while old waves occur at very low wind speeds. This relation is clearly depicted in this figure. Above a wind speed of about 18 m/s no more old waves can be observed in the presented data set.

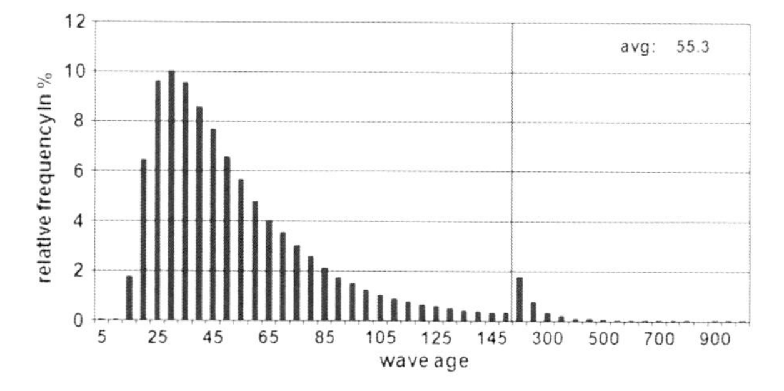

Fig. 5.11 Frequency distribution of wave age at FINO1 for the year 2005. The *vertical line* indicates a breakpoint in bin width: to the *left*: bin width = 5, to the *right*: bin width = 50

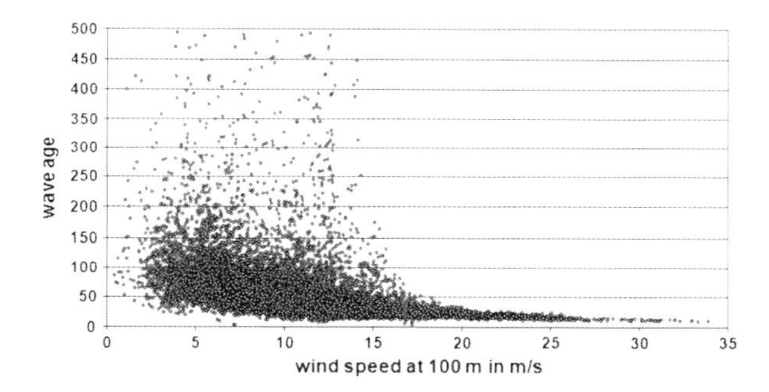

Fig. 5.12 Wave age plotted versus 100 m mean wind speed at FINO1 for the year 2005

5.1.5 Impact of the Vertical Moisture Profile

Furthermore, the perfect moisture source of the sea surface has some additional consequences. While turbulent heat and moisture fluxes are strongly correlated at onshore sites, they are quite often uncorrelated at offshore sites. Turbulent heat fluxes in the marine surface layer depend on the air-sea temperature difference with upward fluxes when the sea is warmer than the air above. Turbulent moisture fluxes are nearly always directed upward, because we nearly always have drier air above more humid air directly over the sea surface. Since humid air is slightly lighter than dry air (for a given temperature), these upward humidity fluxes always contribute to a slight destabilization of the marine surface layer (Sempreviva and Gryning 1996). Oost et al. (2000) also detected negative humidity fluxes together with positive temperature fluxes in the MABL, which they could not explain with classical Monin–Obukhov similarity. Edson et al. (2004) state that "in fact, the moisture flux component… provided more than half of the total buoyancy flux…, and this component kept the surface layer slightly unstable."Recently Barthelmie et al. (2010) estimated that neglecting the humidity influence may lead to an overestimation of the extrapolated mean wind at 150 m from low-level wind speeds by about 4 %.

5.1.6 Annual and Diurnal Variations

The thermal properties of the sea surface are considerably different from those of a land surface. Water has a much larger heat capacity than soil. Therefore, sea surface temperature does not show diurnal temperature variations but mainly an annual variation with a maximum in late summer and a minimum in late winter. This annual variation is slightly modified by cold and warm air advections occurring with moving atmospheric pressure systems on a temporal scale of a few days. Thus, the strong diurnal variation of the vertical structure of the atmospheric boundary layer, which is so familiar from land sites, is completely missing in the marine boundary layer, except for coastal regions when the wind blows from the land (see Sect. 5.6). We rather find a dominant annual variation. Unstable marine boundary layers prevail in autumn and early winter and stable marine boundary layers in spring and early summer. This seasonal pattern comes from the generally larger thermal inertia of the sea water which leads to a time shift in the order of 1 month of the annual temperature variation of the water with respect to the atmospheric annual temperature variation. Therefore, we find cooler air masses over the warmer sea water in autumn while we have warmer air masses over the cool oceans in spring.

5.2 Vertical Profiles

Usually, hub heights in offshore wind parks are above the often quite shallow constant flux or surface layer (see Fig. 5.1). Hub heights are rather in the Ekman layer of the MABL where we find only a slight wind speed increase and a slight turning of the wind direction with height. Therefore, a vertical extrapolation of the wind profiles using the power law (3.22) instead of the stability-dependent logarithmic law (3.16) is suitable. It is demonstrated in Sect. 3.1.3 above that for very smooth surfaces such as the sea surface the difference between the logarithmic profile and the power law profile are small.

Figure 5.13 shows the frequency distribution for the power law profile exponent a from the mast FINO1 in the German Bight. These exponents have been derived from 10 min-averaged wind profiles at the height range between 40 and 90 m taking 40 m as reference height. The most frequent value is 0.03, the mean value is 0.10. These values are much lower than those over land (see Fig. 3.4). The exponent depends considerably on wind speed and thermal stratification (Fig. 5.14). The increase with wind speed is absent over land where the exponent approaches a constant value for very high wind speeds. The offshore power law exponent increases with growing wind speed, because the waves grow and the sea surface gets rougher with increasing wind speed. In the same manner as over land,

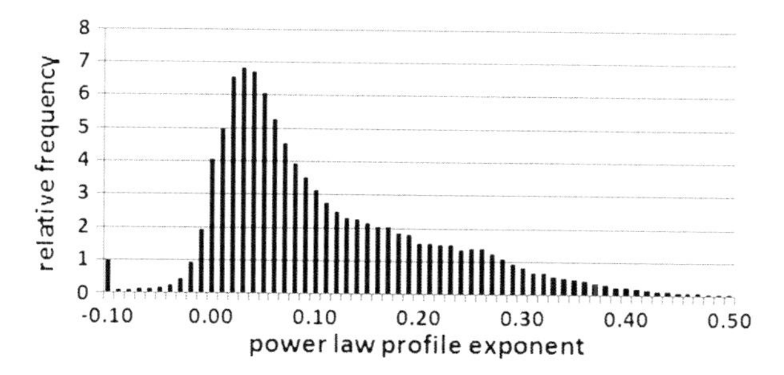

Fig. 5.13 Frequency distribution of the power law profile exponent a in percent [see Eq. (3.22)] for a reference height $z_r = 40$ m at FINO1 in the German Bight for the period September 2003 to August 2007 for wind speeds higher than 5 m/s at 100 m. Bin width is 0.01. The leftmost column summarizes all occurrences with even more negative values

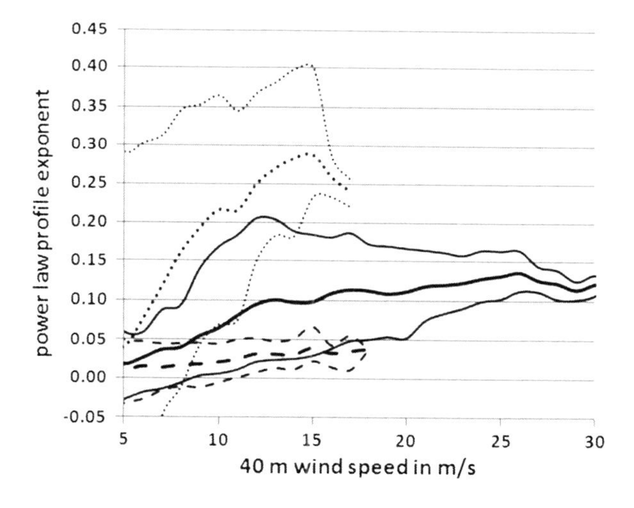

Fig. 5.14 Power law profile exponent a [as in Fig. 5.13)] as function of wind speed at 40 m height at FINO1 in the German Bight for neutral stratification (*bold lines* showing the 10th percentile, the mean and the 90th percentile), unstable stratification (*dashed lines*) and stable stratification (*dotted lines*)

the exponent also grows with increasing thermal stability, because the vertical shear of horizontal wind speed increases in stable conditions due to suppressed vertical turbulent mixing.

While the mean value for the power law profile exponent for neutral conditions is below the value 0.14 which is assumed in the normal wind profile model (NWP) of the standard IEC 61400-3 (2006) for offshore wind turbines, it can happen that the exponent is sometimes above this value (see Fig. 5.14). At mean wind speeds of 12–13 m/s the 90th percentile for the power law exponent is even above the onshore value of 0.20 given in the IEC standard. The 90th percentile decreases again with higher wind speeds while the mean value for this exponent still increases. This is because the distribution of this exponent becomes much narrower with increasing wind speed. For stable stratification the exponent regularly

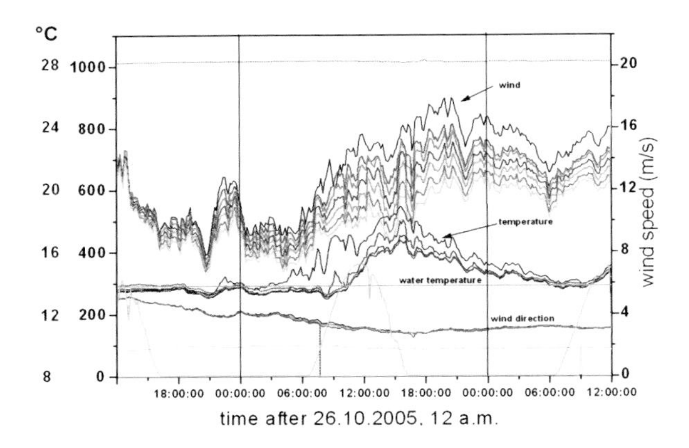

Fig. 5.15 Variation of wind speed at 30, 40, 50, 60, 70. 80, 90, and 100 m height in m/s (*upper* bundle of full curves from *bottom* to *top*, *right* axis), air temperature at 30, 40, 50, 70, and 100 m in °C (*middle* bundle of full curves, from *bottom* to *top*, leftmost axis), wind direction at 30, 50, 70, and 90 m in degrees (*lower* bundle of full curves, *left* axis), sea surface temperature in °C (*horizontal* line labelled "water temperature"), surface pressure in hPa (*upper straight line, left axis*), relative humidity in % (*lower nearly straight line, left axis*) and global radiation in W/m^2 (curve between 0 and 400, *left* axis) at FINO1 in the German Bight for the period October 26, 2005 12 UTC + 1 to October 28, 2005 12 UTC + 1

exceeds the value 0.14 given in the offshore IEC standard. For wind speeds above 15 m/s this even happens for the 10th percentile. For unstable conditions the exponent rarely exceeds a value of 0.05.

Figure 5.15 gives an example how thermal stratification of the MABL directly influences the vertical wind shear. The figure shows a record of 48 h duration. Initially, air temperature is very close to sea surface temperature and the vertical wind shear is small. After about 18 h, an episode of warm air advection starts, which lasts for roughly 24 h. Immediately after the onset of the warm air advection the vertical wind shear increases considerably, visible from the growing spread between the wind speeds at different heights at the mast FINO1. In the afternoon of the second day in the centre of the Figure, 100 m wind speed is about twice as large as 30 m wind speed. This gives a shear of 8 m/s over a height interval of 70 m. This large vertical shear disappears rapidly when the warm air advection ends at the end of the displayed episode. This example shows that the air-water temperature difference is the decisive parameter which governs the vertical shear in the MABL. In contrast to land surfaces, the change in static stability in the MABL is not coupled to the diurnal radiative cycle but to passing weather systems (depressions).

Figure 5.16 gives an example of the monthly distribution of thermal stratification in the MABL by displaying the spread between the potential temperatures at the heights 30 and 100 m for October 2005. Potential temperatures are temperatures corrected for the adiabatic temperature decrease with height. For neutral stratification, potential temperature is constant with height. Potential temperatures increase with height for stable stratification and decrease for unstable stratification. During that month the average sea surface temperature at the mast FINO1 was

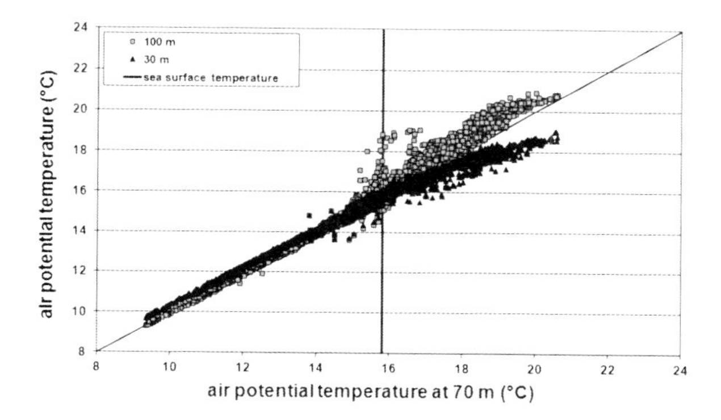

Fig. 5.16 Potential (see text) temperature in °C at 30 m (*full triangles*) and 100 m (*open squares*) plotted against potential temperature at 70 m at FINO1 in the German Bight for October 2005. The *bold vertical line* gives the monthly mean sea surface temperature in °C, the *thin slanted line* gives potential temperature at 70 m

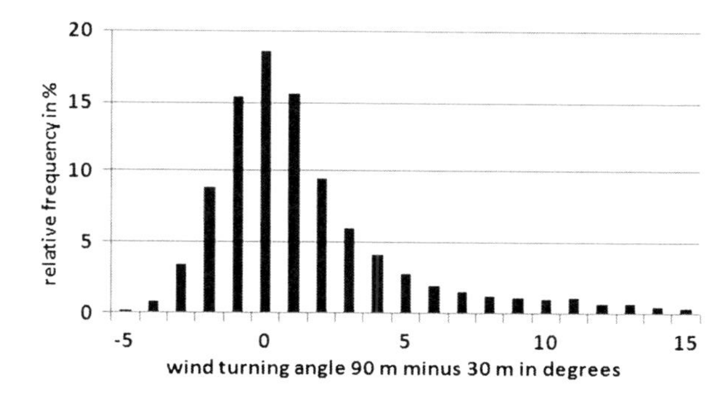

Fig. 5.17 Frequency distribution of the difference in wind direction between 30 and 90 m ($dir_{90\ m}$ minus $dir_{30\ m}$) height at FINO1 in the German Bight for the year 2004

nearly 16 °C, indicated by the vertical line in this figure. Situations with cold air advection are to the left of the vertical line. Here air temperatures were below sea surface temperature, i.e. unstable stratification prevailed. Vertical temperature gradients are small due to the intense thermally induced vertical mixing. Situations with warm air advection are to the right of the vertical line. Here air temperatures are above sea surface temperature and stable stratification is found. Vertical mixing is suppressed and considerable vertical temperature gradients can develop. For an air temperature at 70 m being about 5 °C larger than sea surface temperature the vertical temperature spread between 30 and 100 m grows to about 2 °C. These extreme stable conditions are those where very large power law exponents above 0.30 or even above 0.40 have been found (see Fig. 5.14). The 90th percentile curve for stable stratification in Fig. 5.14 demonstrates that the occurrences of these most extreme shear cases peak at mean wind speeds around 15 m/s.

As already mentioned above, offshore hub heights are usually in the Ekman part of the MABL. This becomes obvious when looking at the wind direction differences between 30 and 90 m height measured at the meteorological mast FINO1

(Fig. 5.17). Although the most frequent turning angle over this height range is around 0°, there are much more positive turning angles (clockwise turning with height) than negative turning angles. Negative values most probably occur with low wind speeds and cold air advection (see last paragraph in Sect. 2.4 on thermal winds above).

5.3 Extreme Wind Speeds

10 min-mean wind conditions were considered in the preceding subchapters. The offshore IEC standard also gives limit values for extreme wind speeds in the extreme wind speed model (EWM). The vertical profiles of 3 s-gusts with a return period of 1 year, v_{e1} and of 50 years, v_{e50} are defined as follows:

$$v_{e1}(z) = 0.8 v_{e50}(z) \tag{5.12}$$

and:

$$v_{e50}(z) = 1.4 v_{ref} \left(\frac{z}{z_{hub}} \right)^{0.11} \tag{5.13}$$

The reference velocity, v_{ref} is put to 50 m/s for class I offshore sites and to 42.5 m/s for class II sites. The vertical profile of 10 min-mean wind speeds with a return period of 1 year, v_1 and of 50 years, v_{50} are defined as follows:

$$v_1(z) = 0.8 v_{50}(z) \tag{5.14}$$

and:

$$v_{50}(z) = v_{ref} \left(\frac{z}{z_{hub}} \right)^{0.11} \tag{5.15}$$

The difference between Eqs. (5.13) and (5.15) is the gust factor 1.4 in (5.13).

Jensen and Kristensen (1989) find a gust factor (see Appendix A, Eq. A.33) at 70 m height on the little island of Sprogø in the Great Belt between the Danish Isles of Fyn and Sjælland of:

$$G(3\text{ s}, 10\,\text{min}, 70\text{ m}, 1.5 10^{-3}, v) \approx 1.15 \tag{5.16}$$

which is considerably lower than the factor 1.4 in Eq. (5.13). Abild and Nielsen (1991) give the following simpler relation for the offshore gust factor which does not depend on wind speed:

$$G(z, z_0) \approx 1 + k I_u = 1 + \frac{k}{\ln \frac{z}{z_0}} \tag{5.17}$$

with k about 2.1. Equation (5.17) has obviously been derived from the relation between the turbulence intensity (A.6) and the gust factor given in Eq. (A.35) in the Appendix stipulating the validity of the logarithmic wind profile by using Eq.

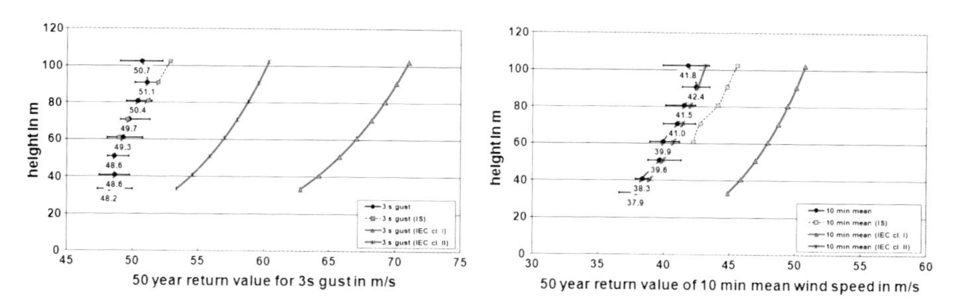

Fig. 5.18 Vertical profiles of extreme 3 s-gusts (*left*) and extreme 10 min-mean wind speeds for 50 year return period. Profiles with *error bars* denote extrapolations from observations at FINO1 using the Gumbel method (see Appendix A.3), *the dashed curve* extrapolations using the independent storm method, the other two curves give the class I (*right*) and class II (*left*) limits from the offshore IEC standard

(3.10). Applying typical values for Sprogø ($z = 70$ m, $z_0 = 0.0015$ m) in Eq. (5.17) yields $G = 1.195$, which is again lower than 1.4. Although it should be paid attention to the fact that (5.17) is simplified, because the wind speed dependence of the roughness length is not considered in this relation, it turns out from relation (5.17) that a gust factor of 1.4 seems to be too high for marine conditions. We will see in the next subsection that offshore turbulence intensity is typically between 0.05 and 0.10 which gives the gust factor from the first relation in (5.17) to be approximately in the range between 1.105 and 1.21.

Figure 5.18 compares vertical profiles of the estimated extreme values for the 50 year return period from FINO1 observations with the offshore IEC standards (5.13) and (5.15). The estimations have been determined for each height separately from 4 years of data. The independent storm method (Cook 1982; Palutikof et al. 1999) has been used to validate the Gumbel method (described in Appendix A.3) employed for the estimation of the extreme values. The Gumbel method estimations from the observations are well below the IEC limit curves for the 3 s-gust but hit the limit curve for class II sites for 10 min-mean wind speeds.

5.4 Turbulence

This subchapter analyses several turbulence parameters which have relevance for load calculations for wind turbines. Most of them are used in the definition of the turbulence models and the extreme operating gust in the IEC standards (IEC 61400-1 Ed. 3 and IEC 61400-3 Ed. 1). These parameters comprise the turbulence intensity, high-frequency wind speed variances, turbulence length scales and inclination angles, and the wind speed variation with time during typical gust events ('Mexican hat').

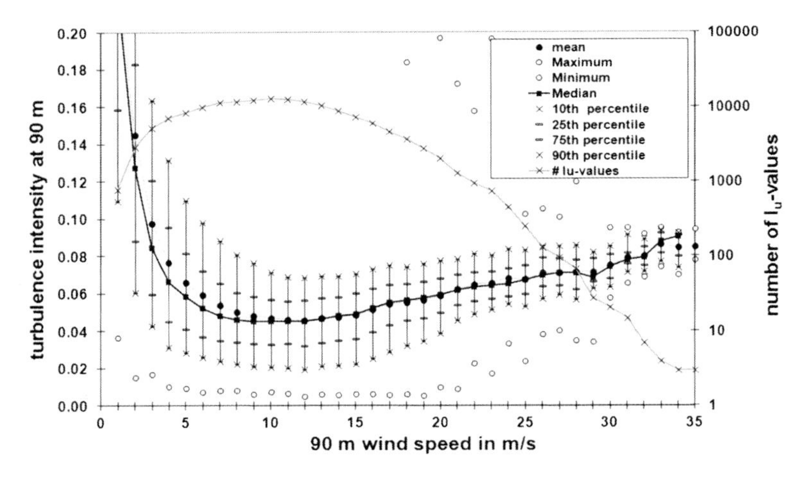

Fig. 5.19 Mean (*full circles*), maximum (*upper open circles*), minimum (*lower open circles*), median (*full line with squares*), 10th percentiles (*lower crosses*), 25th percentiles (*lower dashes*), 75th percentiles (*upper dashes*) and 90th percentiles (*upper crosses*) of turbulence intensity I_u as function of wind speed at 90 m height at FINO1 in the German Bight for the period September 2003 to August 2007 together with the number of values per wind speed bin (crosses connected by a *thin* line) (from Türk and Emeis 2010)

5.4.1 Turbulence Intensity

Offshore turbulence intensity I_u [see definition (A.6) in the appendix] depends on roughness length according to Eq. (3.10) and is therefore a function of wind speed (Hedde and Durand 1994; Vickers and Mahrt 1997). The knowledge of the turbulence intensity over the open sea is relevant for a better general understanding of the marine boundary layer as well as for the construction and operation of offshore wind turbines. Loads on the structure of the turbines and power output both increase with increasing turbulence intensity.

The dependence of median, arithmetic mean, minimum, maximum, the 10th, 25th, 75th, and 90th percentiles of turbulence intensity on wind speed—for the measuring period from September 2003 to August 2007 and a measuring height of 90 m—is shown in Fig. 5.19. For low wind speeds the mean of turbulence intensity rapidly decreases with increasing wind speed to a minimum value of about 4.5 % at 12 m/s wind speed. Above this minimum, turbulence intensity increases nearly linear with increasing wind speed. The high turbulence intensity values and the strong decrease up to wind speeds of about 12 m/s originated from the dominance of thermal induced turbulence at low wind speeds during unstable atmospheric conditions with water surface temperatures significantly above the air temperature. With furthermore increasing wind speed and so increasing roughness length, z_0 the mechanically part of the turbulence intensity begins to dominate over

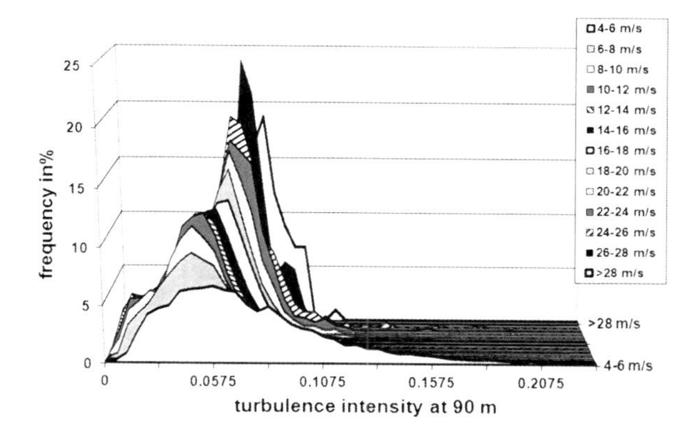

Fig. 5.20 Frequency distributions of turbulence intensity [see Eq. (3.10)] at 90 m height at FINO1 in the German Bight for 13 wind speed classes for the period September 2003 to August 2007

the thermal effects and turbulence intensity increases again (Barthelmie 1999). Maximum values of turbulence intensity at wind speeds below 18 m/s are higher than 20 % and therefore not visible in the plotted range of the turbulence intensity values in Fig. 5.19. They peak at 48.8 % for 13 m/s wind speed and are between 36.3 and 50.0 % for wind speeds between 1 and 13 m/s.

Due to the non-Gaussian frequency distribution median and arithmetic mean of turbulence intensity differ below wind speed values of about 11 m/s while at higher winds speeds these two values are nearly equal. The absolute minimum of turbulence intensity for each 1 m/s wind speed bin is less than 1 % up to wind speeds of 20 m/s. Inspection of the synoptic conditions suggests that under very stable atmospheric conditions situations can occur with very low turbulence intensity even at relative high wind speeds. Above 20 m/s wind speed the influence of the more and more rough surface and so increasing friction stress can always break up this very stable layering and so also the absolute minimum of turbulence intensity begins to increase to values near the 10th percentiles. At higher wind speeds the spread of turbulence intensity values within one wind speed class is continuously becoming smaller (Large and Pond 1981).

Turbulence in the atmospheric boundary layer is either generated by shear or by thermal instability. While for lower wind speeds thermal production of turbulence is dominant it becomes nearly negligible for high wind speeds when compared to shear production. The shear production is proportional to the surface roughness. Onshore the surface roughness is a function of the surface characteristics only and assumed to be independent from the atmospheric conditions. This is different for offshore conditions. The oldest proposition for a description of this dependence is by Charnock (1955) who proposed the relation (5.1). Garratt (1977) reviewed the topic of sea surface roughness and recommended to estimate z_0 by Charnock's relation [Eq. (5.1)] with $\alpha = 0.0144$ for $\kappa = 0.41$ which according to Wu (1980) corresponds to $\alpha = 0.017$ for $\kappa = 0.4$. The IEC standard 61400-3 (IEC 61400-3, 2006) assumes $\alpha = 0.011$ for offshore conditions. Figure 5.20 shows frequency

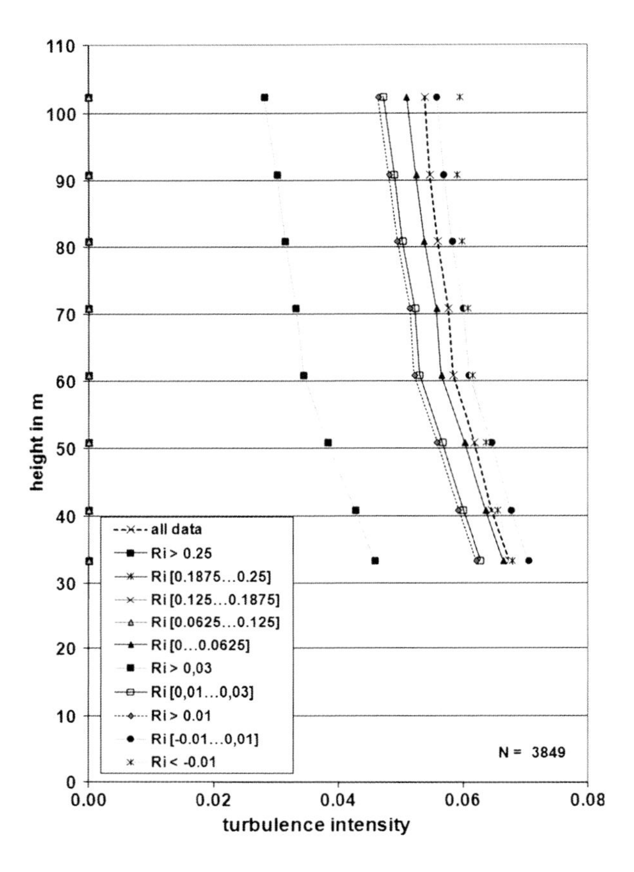

Fig. 5.21 Vertical profiles of turbulence intensity [see Eq. (3.10)] as function of gradient Richardson number at FINO1 in the German Bight for the period October 2004 to January 2005 for the wind direction sector 210°–250°

distributions of turbulence intensity within different wind speed classes at FINO1 at 90 m height from the 4 years of data displayed in Fig. 5.19.

Most measurement heights of the 100 m high FINO1 mast in the German Bight are usually above the well-mixed surface or Prandtl layer. Therefore, we observe a considerable decrease of turbulence with height. This decrease is depicted in Fig. 5.21. This decrease is largest for very stable thermal stratification and is smallest for very unstable stratification.

For the calculation of loads the 90th percentile of the turbulence intensity for a given wind speed bin is important. The normal turbulence model (NTM) of the IEC standards (IEC 61400-1 Ed. 3 and IEC 61400-3 Ed. 1) describe the 90th percentile of the offshore turbulence intensity using the following parameterization for the 90th percentile of the standard deviation of the horizontal wind speed, σ_{u90}:

$$\sigma_{u90} = \frac{u_h}{\ln(z_h/z_0)} + 1.28(1.44\,\text{m/s})I_{15} \tag{5.18}$$

where u_h is the wind speed at hub height of the wind turbines, z_h is the height of the hub above sea level and I_{15} is the average turbulence intensity at hub height at

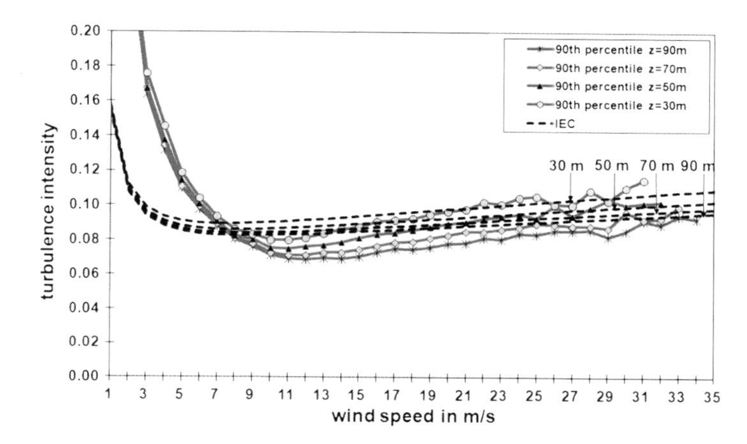

Fig. 5.22 Observed 90th percentiles (*red curves*) of turbulence intensity I_u at four different heights as function of wind speed at these heights (from Türk and Emeis 2010) compared to the results of Eq. (5.18) (*dashed lines*)

15 m/s wind speed. The first term on the right-hand side of (5.18) gives the mean standard deviation of the wind speed for thermally neutral stratification making the assumption of a logarithmic vertical wind profile (3.6) and setting $1/\kappa$ equal to 2.5 [see (3.9)]. For the second term it is assumed that the values for the standard deviation of the wind speed follow a Gaussian distribution around its mean, so that the 90th percentile of the standard deviation of the wind speed, σ_{u90} is 1.28 times the standard deviation of the standard deviation of the wind speed, the latter represented by 1.44 m/s times I_{15}. We will suggest an update to this formula below in Eq. (5.19).

Figure 5.22 shows the 90th percentiles of measured turbulence intensity depending on wind speed at different heights for the period September 2003–August 2007 (solid lines) compared to turbulence intensity given by IEC 61400-3 (Eq. 5.18, dashed lines). Similar to the median of turbulence intensity the values of the 90th percentiles of turbulence intensity also decrease with increasing wind speed till a minimum of about 7–8.5 % at 10–12 m/s wind speed and then increase again with furthermore increasing wind speed. 90th percentiles of turbulence intensity also decrease with height. Compared to turbulence intensities given by the IEC standard, we can detect three sectors: Below wind speeds of 8–10 m/s—at wind speeds that are not really load-relevant—measured 90th percentiles of the turbulence intensities are covered by the IEC-curves not very well. At wind speeds between 10 and 22 m/s the 90th percentiles of measured turbulence intensities lie below the values given by the IEC standard except for two values at 30 m height. In this sector the slopes of the curves of measured turbulence intensity are steeper than the slopes of the curves calculated according to IEC 61400-3. Above wind speeds of about 22 m/s the slopes of measured and calculated curves of the 90th percentile of the turbulence intensity become nearly identical. The discrepancy between the measured data and the computed values from (5.18) for lower wind

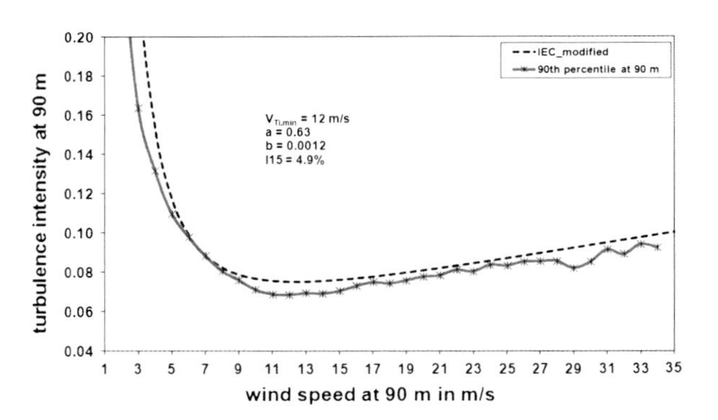

Fig. 5.23 Comparison of the modified IEC relation for turbulence intensity to the 90th percentile of turbulence intensity at 90 m from Eq. (5.19) with FINO1 data

speeds is partly due to the skewed and non-Gaussian distribution of measured values of the standard deviation of the wind speed for wind speeds below 11 m/s (see the discussion of Fig. 5.19). The derivation of (5.18) had assumed a Gaussian distribution for all wind speed bins.

At the upper heights of the FINO1 mast the values according to the standard lie permanently above the measured values, while at the heights of 50 and 30 m the measured values lie above the IEC-values for some wind speed bins. Especially at the measuring height of 30 m influences to turbulence intensity (i.e. an increase of turbulence intensity) from the FINO1 platform structure (for example the helicopter landing deck) cannot be excluded.

The data presented here suggest a modification of relation (5.18). A better fit is possible from:

$$\sigma_{u90} = a\frac{u_h}{\ln(z_h/z_0)} + \frac{2u_{Iu,\min}}{u_h}(1.44\,\mathrm{m/s})I_{15} + bu_h \tag{5.19}$$

where $u_{Iu,\ min}$ is the wind speed at which the minimum turbulence intensity occurs. a and b are two tunable factors. In Fig. 5.23 this better fit is shown where $a = 0.63$ and $b = 0.0012$ have been used.

5.4.2 Wind Speed Variances

In Eq. (3.9), the normalised wind speed-independent standard deviations of the three wind components in the surface layer over flat and homogeneous terrain have been given. In the MABL these values are no longer independent of wind speed, because the surface roughness changes with wind speed. Figure 5.24 shows the normalised variances for the longitudinal, transverse and vertical wind components as function of wind speed at FINO1 in the German Bight. The curves in Fig. 5.24 have to be compared with the squared values from Eq. (3.9) ($\sigma_u^2/u_*^2 = 6.25$, $\sigma_v^2/u_*^2 = 3.61$, $\sigma_w^2/u_*^2 = 1.69$). The normalised variances for 40, 60, and 80 m above the sea surface approach to these values for medium wind speeds for the vertical component and for very high wind speeds for the two horizontal components. For

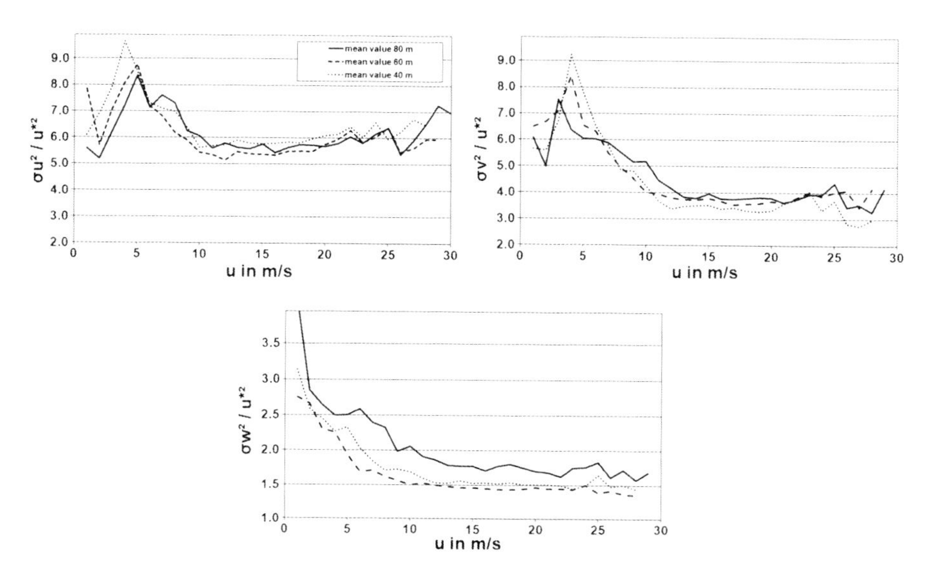

Fig. 5.24 Normalised variances versus wind speed in three different heights (*full line*: 80 m, *dashed line*: 60 m, *dotted line*: 40 m) from sonic data for neutral stratification at FINO1 in the German Bight. *Upper left*: variance of longitudinal component wind component, *upper right*: transverse component, *below*: vertical component

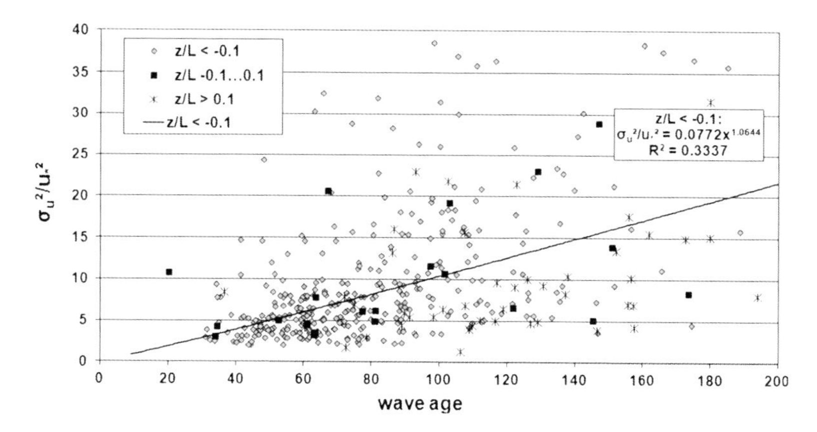

Fig. 5.25 Normalised variance of the longitudinal wind component as function of wave age and thermal stability (*diamonds*: unstable, *squares*: neutral, *crosses*: stable) at 40 m height at FINO1 in the German Bight for the period July to December 2005. Only 40 m mean wind speeds between 3.5 and 4.5 m/s have been considered. The *regression line* is plotted for unstable cases only

medium wind speeds, the values for the horizontal components lie below the values for flat onshore terrain. The three mentioned measurement heights of the FINO1 mast are probably above the well-mixed surface layer for this wind speed range.

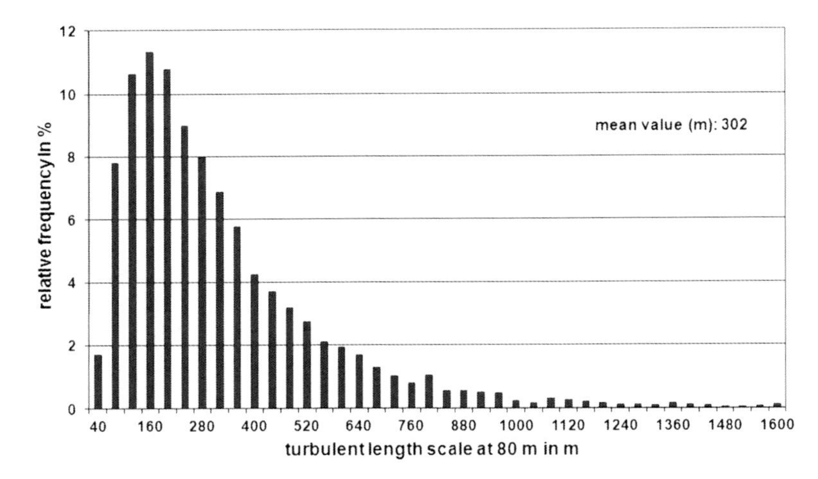

Fig. 5.26 Relative frequency of turbulence length scale in m at 80 m height at FINO1 in the German Bight for the year 2005

Table 5.2 Stability dependent turbulence length scales, Λ_{Smax} at 80 m height at FINO1 determined from maxima in the spectra for the three wind components (taken from Türk 2008)

	$z/L < -1$	$-0,1 < z/L < 0,1$	$z/L > 1$
	Λ_{Smax} (m)	Λ_{Smax} (m)	Λ_{Smax} (m)
u	290	485	292
v	252	286	223
w	223	97	15

The higher values for low wind speeds result from situations with high wave ages (see Sect. 5.1.3). These high wave ages are most frequent with low wind speeds. Figure 5.25 shows that the normalised variance increases with wave age for low wind speeds. Such an increase has been reported by Davidson (1974) as well.

5.4.3 Turbulence Length Scales and Inclination Angles

Another parameter characterizing the spatial scale of the turbulence elements is the turbulence length scale. Turbulence elements which are of similar size as the turbine rotor often hit the rotor only partially and cause differential loads on the rotor. Turbulent length scales have been determined according to the procedure sketched in Sect. A.6 of the Appendix. Offshore turbulence length scales vary from 302 m for the longitudinal component (Fig. 5.26) via 273 m for the transverse component to 41 m for the vertical component. These length scales can also be

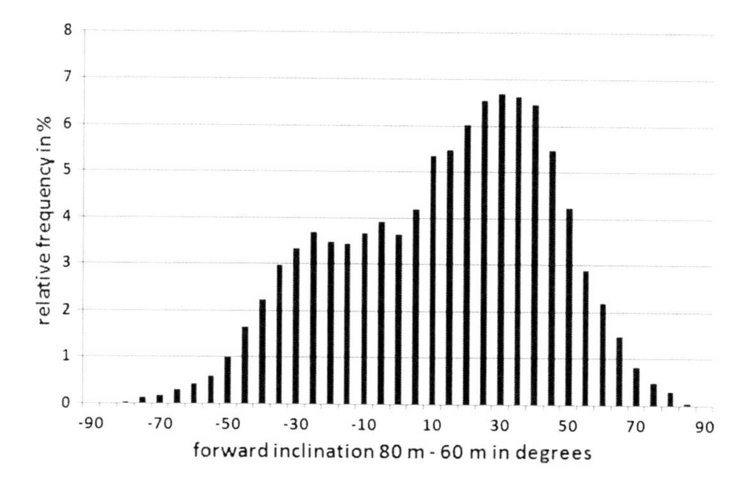

Fig. 5.27 Forward inclination of turbulence elements in degrees between 60 and 80 m height at FINO1 in the year 2005. Negative values indicate backward inclination

determined from maxima in the spectra of the three wind components. This gives values in the same order of magnitude but also shows considerable differences for the three stability classes with the largest values for the horizontal components with neutral stratification and largest values for the vertical component with unstable stratification. Details can be found in Table 5.2.

Another parameter characterizing the turbulence is the inclination of the turbulence elements with respect to the vertical. A forward inclination of the turbulence elements is expected, because the wind speed decreases towards the surface. Forwardly inclined elements show up earlier at greater heights than at lower heights, i.e. the upper tip of a rotor is impacted slightly earlier than the lower tip. This can cause differential loads on the rotor. Figure 5.27 shows that this forward inclination occurs in the majority of all 10 min intervals. An inclination angle of 30° between 60 and 80 m is most frequent while an angle of about 40° is most frequent between 40 and 60 m (not shown) due to the larger vertical shear in that height interval. The inclination increases with wind speed. For wind speeds higher than about 20 m/s nearly no backward inclinations are found any more (see Fig. 5.28). Figure 5.28 also demonstrates that large inclination angles usually coincide with non-neutral thermal stratification.

5.4.4 Gust Events

So far, bulk statistical parameters characterizing atmospheric turbulence have been discussed. But the actual wind speed variation during a gust event can be decisive as well for load calculations. The IEC standard 61400-3 (2006) defines as a worst

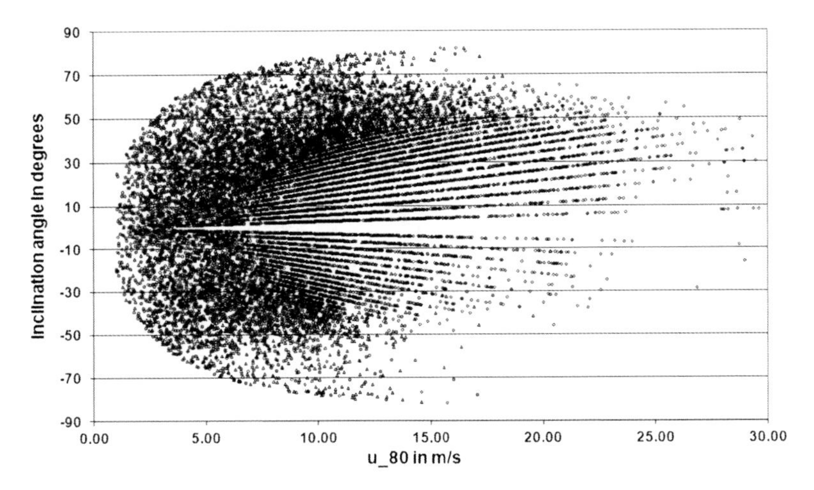

Fig. 5.28 Inclination angle of turbulence elements versus 80 m mean wind (only values larger than 1 m/s have been evaluated here) speed at FINO1 for the year 2005. *Diamonds*: near neutral stratification, *triangles*: stable stratification, *squares*: unstable stratification. Ray patterns in the centre of the figure result from the limited resolution of wind speed data (two decimal digits)

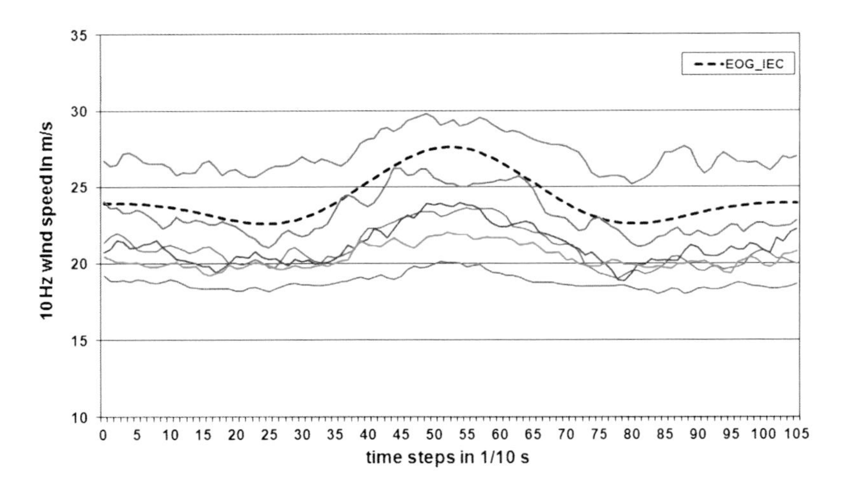

Fig. 5.29 10 Hz wind speed data showing six selected gust events observed at FINO1 (*full curves*). The *dashed line* shows the idealized "Mexican hat" defined as EOG in the IEC standard

case an extreme operating gust (EOG) as an event with a duration of 10.5 s for this purpose. The typical temporal structure of the gust event starts with a decrease of the wind speed followed by a larger increase and a final decrease before the undisturbed wind speed is reached again. Due to its shape the temporal pattern of this EOG is also called "Mexican hat". See Sect. A.5 for details.

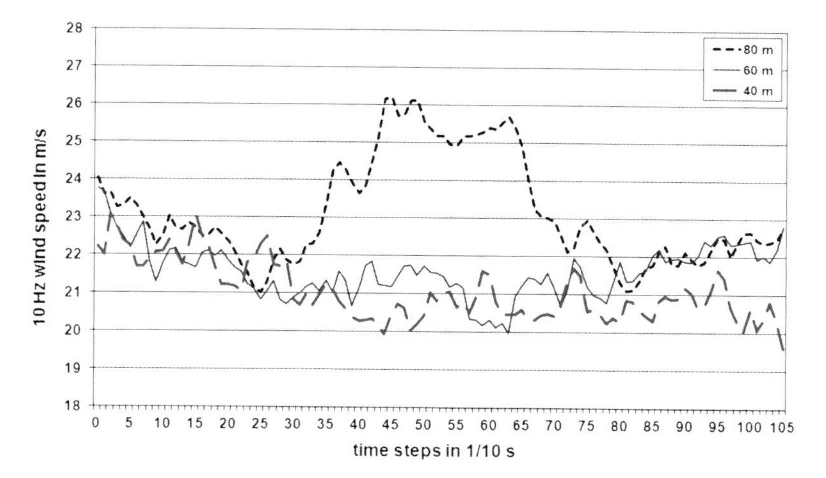

Fig. 5.30 Observed 10.5 s-gust event at FINO1 which was present at 80 m height only. The data start at January 7, 2005, 12.52.32 UTC + 1

Figure 5.29 compares selected, actually measured gust events with the idealiezed definition of an EOG in the IEC standard. The selection procedure analysed the high-resolution wind speed time series and marked all 10.5 s intervals where the IEC EOG explained more than 85 % of the wind speed variance. The procedure did not distinguish between positive and negative correlations. Actually, for 2005, 57 % of all selected gust events had a negative correlation with the EOG, i.e. they were upside down Mexican hats (see also Fig. 5.31). Another remarkable feature was that the gust events did not always appear in all three measurement heights (40, 60, and 80 m). Figure 5.30 shows an example where the gust event was visible at 80 m height only.

Figure 5.31 displays the relative frequency of different amplitudes of these gust events. The mean amplitude is about 12 % for wind speeds below 12 m/s and then slightly increases to about 20 % for wind speeds of 20 m/s and above.

The above evaluations have been made for a 10.5 second-event as defined in the IEC standard. Changing the event duration in the selection procedures showed that 10.5 second-events are not the most frequent events. 8 second-events occur 1.6 times as frequent as 10.5 second-events, while 14 second-events only occur 0.63 times as frequent.

5.5 Weibull Parameter

The measurements at the FINO1 platform in the German Bight allow for an analysis of the variation of the Weibull parameters in a marine boundary layer with height and season [see Bilstein and Emeis (2010) for further details]. Figure 5.32

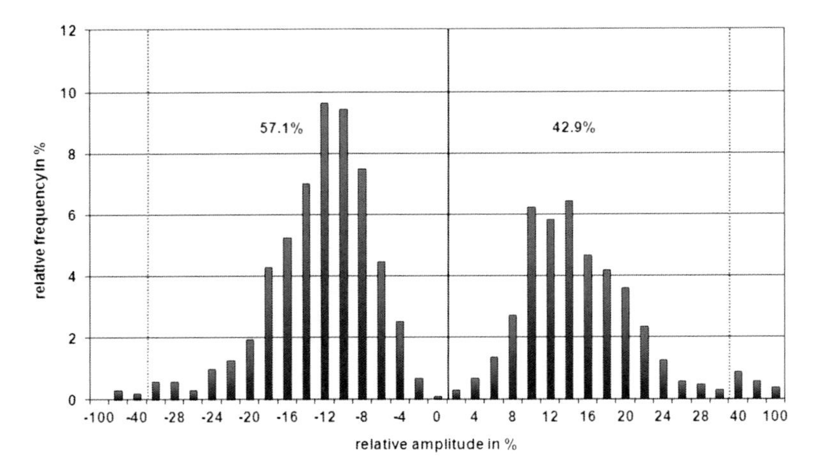

Fig. 5.31 Relative frequency of relative amplitudes (percentage of mean wind speed) of 10.5 s-gust events at 80 m at FINO1 for the year 2005. The *dotted vertical lines* indicate a change of bin width (2 % in the *interior* part of the Figure and 30 % in the *outer* parts)

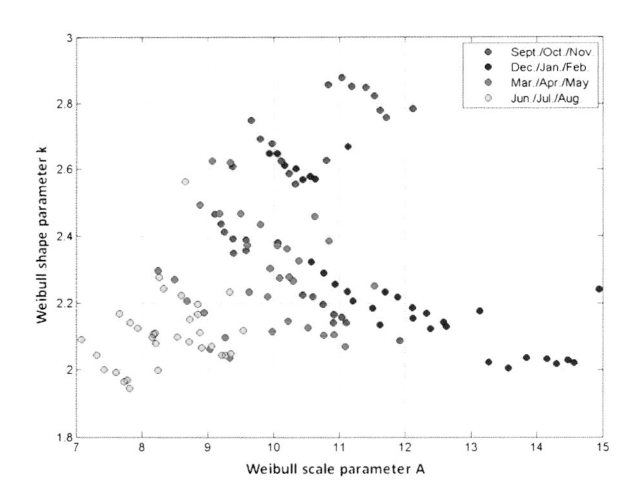

Fig. 5.32 Correlation between the two Weibull parameters A and k at each height at FINO1, where *yellow* are summer, *green* are spring, *red* are autumn and *blue* are wintertime data. Within each seasonal result the different measurement heights order from *upper left* (30 m) to *lower right* (100 m) with 10 m increment. (From Bilstein and Emeis 2010)

shows data for the four different seasons and for all eight heights between 30 and 100 m with a height increment of 10 m. Please note that the upmost instrument at 100 m is on the top of the mast while all other instruments are mounted on horizontal booms away from the mast. These lower instruments are slightly influenced by the mast. This is why the last data point in the lower right of each profile shown in Fig. 5.32 is a bit shifted to the right. It becomes visible that the shape parameter decreases with the rising variability of wind speeds at higher levels, while the scale parameter increases with height. Over the ocean, atmospheric friction is not as great as over land, so the surface layer (also called Prandtl layer) is not as thick (Türk 2008). For this reason, the vertical gradient of the scale

Table 5.3 Mean of the scale parameter of the Weibull distribution A in m/s for all seasons at 90 and 40 m and their vertical difference in m/s

Seasons	90 m	40 m	Difference
SON (autumn)	10.68	10.12	0.56
DJF (winter)	12.36	11.57	0.79
MAM (spring)	10.27	9.23	1.04
JJA (summer)	8.55	8.03	0.52

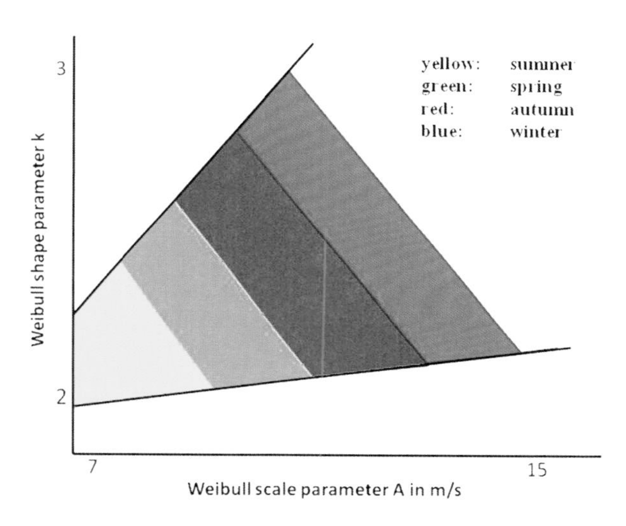

Fig. 5.33 Schematic diagram of the seasonal variation of the correlation of the two Weibull parameters A and k in a marine boundary layer. The colouring is the same as in Fig. 5.32. (From Bilstein and Emeis 2010)

parameter $0.01 < \partial A/\partial z < 0.02\ \mathrm{s}^{-1}$ (see Table 5.3) is not as big as in a similar height over land $0.02 < \partial A/\partial z < 0.04\ \mathrm{s}^{-1}$, (see, e.g., Emeis 2001). The offshore shape parameter profile does not show a maximum as the onshore shape parameter profile does, but decreases monotonically with height.

Furthermore, the two Weibull parameters show a clear seasonal dependence. Smaller parameters are detected in summer, higher parameters in winter, while spring and autumn are between both extremes. This is explained by the annual variation of thermal stability in the marine boundary layer. Due to the enhanced heat capacity of water, the stability of the marine atmosphere is out of phase by about 3 months compared to the stability of the atmosphere over land (Coelingh et al. 1996). Consequently the atmosphere is stable in spring, neutral/stable in summer, unstable in autumn and neutral/unstable in winter.

On the whole those seasons, which are stable or neutral/stable have parameters smaller in magnitude compared with unstable seasons. It is noted, that shape parameters with larger scale parameters (autumn and winter) have a higher variability compared with ones with smaller scale parameters (spring and summer). Figure 5.33 shows a schematic diagram of this correlation, where intervals for each season and parameter are presented.

5.6 Coastal Effects

When the larger-scale winds blow from the land to the sea, then an internal boundary layer (IBL) forms over the sea surface (see Sect. 3.5 for an introduction to internal boundary layers and the description of wind profiles in and above these IBLs). If warmer air flows over colder water, then the thermal stratification is stable and the internal boundary layer is growing slowly in depth and can persist over distances of more than 50 kilometres. Aircraft measurements in a stable IBL over the Irish Sea are presented in Rogers et al. (1995). These measurements show profiles of mean quantities and spectra. The spatial development of stable IBLs is described, e.g., by Mulhearn (1981) and Garratt (1987). They give for the height h(x) of the stably stratified IBL:

$$h(x) = cu\sqrt{\frac{x}{g}}\left(\frac{\Delta T}{T}\right)^{n} \tag{5.20}$$

where x is the distance to the coast, T is air temperature, ΔT is the surface air temperature difference between land and sea and g is gravity. Mulhearn (1981) gives c $= 0.0146$ and $n = -0.47$ while Garratt (1987) gives $c = 0.014$ and $n = -0.5$.

If colder air flows over warmer water, the internal boundary layer grows rapidly in depth and is finally merged into the marine boundary layer after some tens of kilometres. In such offshore flows in coastal regions we can observe the usual diurnal changes in atmospheric temperature, stability and winds, which are well-known from flow over land and which have been described in Chap. 3. Thus, the statements in Sect. 5.1.6 are not valid in coastal regions with offshore winds.

5.6.1 Land and Sea Winds

There are local wind systems which do not emerge from large-scale pressure differences but from regional or local differences in thermal properties of the Earth's surface. These local or regional wind systems often exhibit a large regularity and have a sufficient depth so that they can be used for the energy generation from the wind. See Atkinson (1981) for an overview on thermally induced circulations.

Due to the different thermal inertia of land and sea surfaces, secondary circulation systems—land-sea wind systems—can form at the shores of oceans and larger lakes which modify the ABL structure. Under clear-sky conditions and low to moderate winds, land surfaces become cooler than the adjacent water surface due to long-wave emittance at night and they become warmer than the water surface due to the absorption of short-wave irradiance during daytime. As a consequence, rising motion occurs over the warmer and sinking motion over the cooler surfaces. A flow from the cool surface towards the warm surface develops

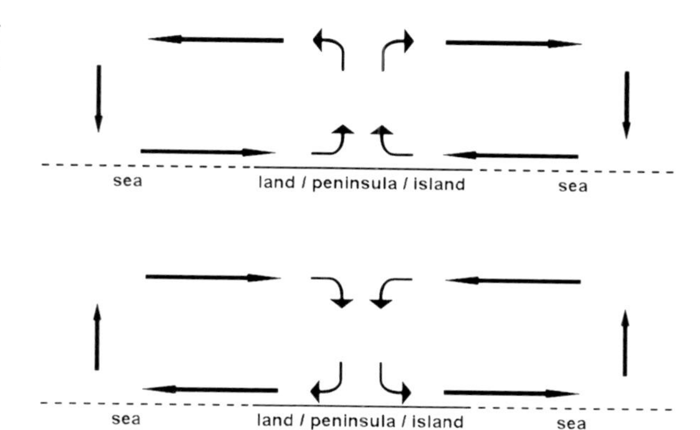

Fig. 5.34 Principal sketch of a land-sea wind circulation at daytime (*top*) and at night-time (*below*)

near the surface and a return flow emerges in the opposite direction aloft in order to keep the mass balanced.

This leads to the well-known sea breeze during daytime and in the evening and the land breeze at night and in the early morning (Fig. 5.34). This common feature of a land breeze is the reason why in former times sailing ships left harbours in the early morning and tended to return in the afternoon. The sea breeze front propagates inland several tens of kilometres during the day and is—if enough moisture is available in the air—often marked by a chain of cumulus clouds. In mid-latitudes sea breezes tend to penetrate 20–50 km but in the tropics distances of up to 300 km and over have been observed. The depth of this flow ranges from a few hundreds of metres to one to two kilometres. Maximum wind speeds in sea breezes can be around 10–11 m/s at about 100 m height (Atkinson 1981). Sea breezes originate from a 100 to 120 km broad coastal zone over the water, detectable from satellite images showing cloud-free conditions in this space (Simpson 1994). The clouds are dissolved due to the sinking motion in this marine branch of the sea breeze (see outer downward arrows in the upper frame of Fig. 5.34). Fewer observations are available for the nocturnal land breeze, but it can be assumed that the spatial extent of these winds is comparable to the extent of the sea breeze.

5.6.2 Low-Level Jets

The offshore vertical wind profile is not always a monotonically increasing function with height. Sometimes wind speed maxima occur within or at the top of the boundary layer. The formation of low-level jets over flat terrain requires a diurnal variation in the thermal stratification of the surface layer. For details refer to Sect. 3.4.2. Although the necessary conditions for the formation of low-level jets are usually absent over ocean surfaces, we sometimes observe low-level jets in the marine boundary layer as well. For example, there are frequent reports from the Baltic Sea and other coastal areas, especially in times when the sea surface is considerably colder than the surrounding land surface and the wind is directed from the land to the sea (Smedman et al. 1995). Brooks and Rogers (2000) have observed low-level jets over the Persian Gulf as well.

The jets can form at distances of several tens of kilometres from the coastline when warm air is advected over the colder water surface. Most likely, such marine low-level jets are spatial analogues to the nocturnal low-level jets over land, which occur due to temporal changes in surface layer stratification. When the air flow passes the coastline and reaches the colder water surface, an internal boundary layer forms and the same decoupling between the surface layer and the rest of the boundary layer above takes place which happens in the evening when the ground cools down due to radiative energy losses.

5.7 Summary for Marine Boundary Layers

The marine atmospheric boundary layer (MABL) shows remarkable differences compared to the atmospheric boundary layer (ABL) over flat homogeneous terrain. Wind speeds are higher and turbulence intensities in the MABL are lower than in the same height over the surface in the onshore ABL (given the same synoptic forcing through a large-scale horizontal pressure gradient). Vertical wind shear over the rotor area of a modern large 5 MW wind turbine is in most cases considerably lower for offshore conditions than over land. In addition, all these MABL characteristics are not constant but vary with wind speed because the roughness of the sea surface increases with wind speed. A further complication may arise from the wave age, i.e. the ratio between the wind speed and the phase speed of the waves. Therefore, the actual wind conditions always depend on atmospheric conditions and on the properties of the wave field.

Wind shear due to thermal winds (see Sect. 2.4 for an estimation of the size of thermal wind) has to be taken into account over the seas because the wind shear due to surface friction is rather low. Usually colder air masses coincide with low-pressure areas and warmer air masses with high-pressure areas in temperate latitudes. Therefore, thermal winds usually contribute to an increase of wind speed with height.

In coastal areas up to about 50–100 km away from the coast, internal boundary layers (see Sect. 3.5) and low-level jets at the top of these layers (Sect. 3.5.1) may occur when the wind blows from the land. The top height of these internal boundary layers in the coastal MABL is often within the rotor area.

The low level of turbulence in the MABL is advantageous for single wind turbines, because it leads to reduced loads on the structure of the wind turbine. But the low turbulence level may turn into a major disadvantage for the planning and operation of larger offshore wind parks, because the supply of kinetic energy towards large wind parks by vertical turbulent fluxes is considerably lower in the ABL. This will be explained in more detail in the following chapter.

References

Abild, J., B. Nielsen: Extreme values of wind speeds in Denmark. Risoe-M-2842, 106 pp. (1991)

Anderson, R.: A study of wind stress and heat flux over the open ocean by the inertial-dissipation method. J. Phys. Oceanogr. **23**, 2153–2161 (1993)

Atkinson B.W.: Meso-scale Atmospheric Circulations. Academic Press, London etc., 495 pp. (1981)

Barthelmie, R.J., A.M. Sempreviva, S.C. Pryor: The influence of humidity fluxes on offshore wind speed profiles. Ann. Geophys. **28**, 1043–1052 (2010)

Barthelmie R.J.: Monitoring Offshore Wind and Turbulence Characteristics in Denmark, Proceedings of the BWEA Wind Energy Conference (1999)

Bilstein, M., S. Emeis: The Annual Variation of Vertical Profiles of Weibull Parameters and their Applicability for Wind Energy Potential Estimation. DEWI Mag. **36**, 44–50 (2010)

Black, P., and Coauthors: Air–sea exchange in hurricanes: Synthesis of observations from the Coupled Boundary Layer Air–Sea Transfer experiment. Bull. Amer. Meteor. Soc. **88**, 357–374 (2007)

Brooks, I., D. Rogers: Aircraft observations of the mean and turbulent structure of a shallow boundary layer over the Persian Gulf. Bound.-Lay. Meteorol. **95**, 189–210 (2000)

Bye, J.A.T., Wolff, J.-O.: Charnock dynamics: a model for the velocity structure in the wave boundary layer of the air–sea interface. Ocean Dyn. **58**, 31–42 (2008)

Charnock, H.: Wind stress on a water surface. Quart. J. Roy. Meteor. Soc. **81**, 639-640 (1955)

Coelingh, J., van Wijk, A., Holtslag, A.: Analysis of wind speed observations over the North Sea. J. Wind Eng. Ind. Aerodyn. **61**, 51–69 (1996)

Cook, N.J.: Towards better estimation of extreme winds, J. Wind Eng. Ind. Aerodyn. **9**, 295–323 (1982)

Davidson, K. L.: Observational Results on the Influence of Stability and Wind-Wave Coupling on Momentum Transfer Fluctuations over Ocean Waves. Bound.-Lay. Meteorol. **6**, 305–331 (1974)

Donelan, M. A.: Air–sea interaction. The Sea. Ocean Eng. Sci. **9**, 239–292 (1990)

Donelan, M.A., B. Haus, N. Reul, W. Plant, M. Stiassnie, H. Graber, O. Brown, E. Saltzman: On the limiting aerodynamic roughness of the ocean in very strong winds. Geophys. Res. Lett. **31**, L18306, doi:10.1029/2004GL019460 (2004)

Edson, J.B., C.J. Zappa, J.A. Ware, W.R. McGillis, J.E. Hare: Scalar flux profile relationships over the open ocean. J. Geophys. Res. **109**, C08S09, doi:10.1029/2003JC001960 (2004)

Emeis, S., M. Türk: Wind-driven wave heights in the German Bight. Ocean Dyn. **59**, 463–475. (2009)

Emeis, S.: Vertical variation of frequency distributions of wind speed in and above the surface layer observed by sodar. Meteorol. Z. **10**, 141-149 (2001)

Fairall, C. W., E. F. Bradley, D. P. Rogers, J. B. Edson, G. S. Young: Bulk parameterization of air-sea fluxes for Tropical Ocean-Global Atmosphere Coupled-Ocean Atmosphere Response Experiment. J. Geophys. Res. **101**, 3747–3764 (1996)

Fairall, C. W., E. F. Bradley, J. E. Hare, A. A. Grachev, J. B. Edson: Bulk parameterization of air–sea fluxes: Updates and verification for the COARE algorithm. J. Climate, **16**, 571–591 (2003)

Foreman, R., S. Emeis: Revisiting the Definition of the Drag Coefficient in the Marine Atmospheric Boundary Layer. J. Phys. Oceanogr. **40**, 2325–2332 (2010)

Garratt, J.R.: Review of Drag Coefficients over Oceans and Continents, Mon. Wea. Rev. **105**, 915–929 (1977)

Garratt, J.R.: The stably stratified internal boundary layer for steady and diurnally varying offshore flow. Bound.-Lay. Meteorol. **38**, 369–394 (1987)

Geernaert, G.: Bulk parameterizations for the wind stress and heat fluxes. Surface Waves and Fluxes, KluwerAcademic, 91–172 (1990)

Hedde, T., Durand, P.: Turbulence Intensities and Bulk Coefficients in the Surface Layer above the Sea. Bound.-Lay. Meteorol. **71**, 415–432 (1994)
Hersbach, H., Janssen, P.A.E.M.: Improvement of the short-fetch behavior in the wave ocean model (WAM). J. Atmos. Ocean Technol. **16**, 884–892 (1999)
Hsu, S.: A dynamic roughness equation and its application to wind stress determination at the air-sea interface. J. Phys. Oceanogr. **4**, 116–120 (1974)
Janssen, J.A.M.: Does the wind stress depend on sea-state or not? A statistical error analysis of HEXMAX data. Bound.-Lay. Meteor. **83**, 479–503 (1997)
Jensen, N.O., L. Kristensen: Gust statistics for the Great Belt Region. Risoe-M-2828, 21 pp. (1989)
Kumar, V.S., Deo, M.C., Anand, N.M., Chandramohan, P.: Estimation of wave directional spreading in shallow water. Ocean Eng. **26**, 23–98 (1999)
Large, W.G., Pond, S.: Open Ocean Momentum Flux Measurements in Moderate to Strong Winds. J. Phys. Ocean. **11**, 324–336 (1981)
Maat, N., C. Kraan, W. Oost: The roughness of wind waves. Bound.-Layer Meteor. 54, 89–103 (1991)
Mulhearn, P.: On the formation of a stably stratified internal boundary layer by advection of warm air over a colder sea. Bound.-Lay. Meteorol. **21**, 247–254 (1981)
Neumann, G.: On ocean wave spectra and a new method of forecasting wind-generated sea. Beach Erosion Board, Washington. Tech. Mem. no. 43 (Dec) (1953)
Oost, W.A., C.M.J. Jacobs, C. van Oort: Stability effects on heat and moisture fluxes at sea. Bound.-Lay. Meteorol. **95**, 271–302 (2000)
Oost, W.A., Komen, G.J., Jacobs, C.M.J., Van Oort, C.: New evidence for a relation between wind stress and wave age from measurements during ASGAMAGE. Bound.-Lay. Meteorol. **103**, 409–438 (2002)
Palutikof, J.P., B.B. Brabson, D.H. Lister, S.T. Adcock: A review of methods to calculate extreme wind speeds. Meteorological Applications, **6**, 119–132 (1999)
Rogers, D.P., D.W. Johnson, C.A. Friehe: The Stable Internal Boundary Layer over a Coastal Sea. Part I: Airborne Measurements of the Mean and Turbulent Structure. J. Atmos. Sci. **52**, 667–683 (1995)
Roll, H.U. : Über Größenunterschiede der Meereswellen bei Warm-und Kaltluft. Dtsch Hydrogr. Z. **5**, 111–114. (1952)
Sempreviva, A.M., S.-E. Gryning: Humidity fluctuations in the marine boundary layer measured at a coastal site with an infrared humidity sensor. Bound.-Lay. Meteorol. **77**, 331–352 (1996)
Simpson, J.E.: Sea breeze and local wind. Cambridge University Press, Cambridge (UK), 239 pp. (1994)
Sjöblom, A., Smedman, A.-S.: Vertical structure in the marine atmospheric boundary layer and its implication for the internal dissipation method. Bound.-Lay. Meteorol. **109**, 1–25 (2003)
Smedman, A.-S., H. Bergström, U. Högström: Spectra, Variances and Length Scales in a Marine Stable Boundary Layer Dominated by a Low Level Jet. Bound.- Lay. Meteorol. **76**, 211–232 (1995)
Smith, S., and Coauthors: Sea surface wind stress and drag coefficients: The HEXOS results. Bound.-Layer Meteor. **60**, 109–142 (1992)
Smith, S.D.: Wind Stress and Heat Flux over the Ocean in Gale Force Winds. J. Phys. Ocean. **10**, 709–726 (1980)
Sullivan, P., J. McWilliams: Dynamics of winds and currents coupled to surface waves. Annu. Rev. Fluid Mech. **42**, 19–42 (2010)
Sverdrup, H.U., Munk, W.H.: Wind, sea and swell: Theory of relations for forecasting. Hydrogr. Off. Publ., No. 601 (1947)
Toba, Y.: Stochastic form of the growth of wind waves in a single parameter representation with physical implications. J. Phys. Oceanogr. **8**, 494–507 (1978)
Türk, M., S. Emeis: The dependence of offshore turbulence intensity on wind speed. J. Wind Eng. Ind. Aerodyn. **98**, 466–471 (2010)

Türk, M.: Ermittlung designrelevanter Belastungsparameter für Offshore-Windkraftanlagen. PhD thesis University of Cologne (2008) (Available from: http://kups.ub.uni-koeln.de/2799/)
Vickers, D., Mahrt, L.: Fetch limited Dr ag Coefficients. Bound.-Lay. Meteorol. **85**, 53–79 (1997)
Wu, J.: Wind-Stress Coefficients over Sea Surface near Neutral Conditions – A Revisit. J. Phys. Oceanogr. **10**, 727–740 (1980)
Yelland, M.J., B. Moat, P. Taylor, R. Pascal, J. Hutchings, V. Cornell: Wind stress measurements from the open ocean corrected for airflow distortion by the ship. J. Phys. Oceanogr. **28**, 1511–1526 (1998)

Chapter 6
Physics of Wind Parks

Wind parks need special treatment, because here the flow conditions approaching most of the turbines in the park interior are no longer undisturbed. Wakes produced by upwind turbines can massively influence downwind turbines. This includes reduced wind speeds and enhanced levels of turbulence which will lead to reduced yields and enhanced loads. For a given land or sea area, it is desirable to place the wind turbines as close together as possible to maximize energy production. However, if wind turbines are too closely spaced, wake interference effects could result in a considerable reduction in the efficiency of the wind park's energy production. Some wind parks with tightly spaced turbines have produced substantially less energy than expected based on wind resource assessments. In some densely packed parks where turbines have failed prematurely, it has been suspected that these failures might have been caused by excessive turbulence associated with wake effects (Elliot 1991).

A special spatial arrangement of the turbines in smaller wind parks with regard to the mean wind direction may help to minimize wake-turbine interactions. But for larger wind parks, wake-turbine interactions are unavoidable in the park interior and the ratio between mean turbine distance and rotor diameter becomes the main parameter that governs the park efficiency. Before we consider such large wind parks in Sect. 6.2, we will shortly describe the characteristics of single turbine wakes.

6.1 Turbine Wakes

We distinguish between near wake and far wake when looking at turbine wakes. The near wake is taken as the area just behind the rotor, where the special properties of the rotor itself can still be discriminated, so approximately up to a few rotor diameters downstream. We find features such as 3D vortices and tip vortices from single blades in the near wake. The presence of the rotor is apparent by the

S. Emeis, *Wind Energy Meteorology*, Green Energy and Technology,
DOI: 10.1007/978-3-642-30523-8_6, © Springer-Verlag Berlin Heidelberg 2013

number of blades, and blade aerodynamics. The far wake is the region beyond the near wake, where modelling the actual rotor is less important (Vermeer et al 2003).

The wake velocity deficit, the downwind decay rate of the wake, and the added turbulence intensity within the far wake with respect to downwind distance behind wind turbines are largely determined by two factors: the turbine's thrust coefficient [see Eq. (6.13) and Fig. 6.2] and the ambient atmospheric turbulence [often characterized by the parameter 'turbulence intensity', see Eq. (3.10)]. The initial velocity deficit depends on the amount of momentum extracted by the turbine from the ambient flow. Thus, this deficit is a function of the turbine's thrust coefficient. Turbine thrust coefficients are generally highest at low wind speeds around the cut-in wind speed and decrease with increasing wind speed. They approach to very low values above the rated wind speed of the turbine. Nevertheless, published data on wake deficits have often been analyzed as a function of wind speed rather than thrust coefficient. Wake measurement data generally verify that deficits are highest at low wind speeds and lowest at high wind speeds (Elliot 1991). Vermeer et al (2003) give the following relation for the distance-dependent relative velocity deficit in the far wake:

$$\frac{\Delta u}{u_h} = \frac{u_{h0} - u_h}{u_h} = A\left(\frac{D}{s}\right)^n \tag{6.1}$$

where u_h is the wind speed at hub height, D is the rotor diameter, s is the distance from the turbine, and A and n are constants. A depends on the turbine thrust coefficient and increases with it. A varies between 1 and 3 while n takes values between 0.75 and 1.25 and principally depends on the ambient turbulence intensity. The WAsP model (Troen and Petersen 1989) uses a similar approach (Barthelmie and Jensen 2010):

$$\frac{u_h}{u_{h0}} = \left(1 - \sqrt{1 - C_t}\right)\left(\frac{D}{D + ks}\right)^2 \tag{6.2}$$

with the turbine thrust coefficient C_t (see (6.13) and Fig. 6.2) and the wake decay coefficient k. $k = 0.04$ is typical for offshore conditions (Barthelmie and Jensen 2010) while 0.075 is the default value in WAsP (Barthelmie et al (2004).

The added turbulence intensity in the wake decreases more slowly than the velocity deficit. Vermeer et al (2003) give three empirical formulae from three different sources which describe the measured data quite well. According to Quarton (1989) the added turbulence intensity decreases as:

$$\Delta I = \sqrt{I^2 - I_\infty^2} = 4.8 C_T^{0.7} I_\infty^{0.68} \left(\frac{s_N}{s}\right)^{0.57} \tag{6.3}$$

where I_∞ is the undisturbed turbulence intensity, C_T is the thrust coefficient, and s_N is the length of the near wake which is between one and three rotor diameters. The width of the wake is proportional to the one third power of the rotor diameter (see Frandsen et al 2006 for more details):

$$D_W(x) \propto s^{1/3} \tag{6.4}$$

This spreading of the wake with distance downstream of the turbine leads unavoidably to complex wake–wake interactions in larger wind parks. The formulation for multiple wakes in WAsP is a quadratic superposition of the single wakes (bottom-up approach, Barthelmie and Jensen 2010)

$$\left(1 - \frac{u_h}{u_{h0}}\right)^2 = \sum_n \left(1 - \frac{u_{hn}}{u_{h0}}\right)^2 \tag{6.5}$$

with $n = 1,\ldots N$ the contributions from N single wakes. Jensen (1983) derived for an infinite number of turbines in a row the following asymptotic expression:

$$\frac{u_h}{u_{h0}} = 1 - \left(\frac{a}{1-a}\right)\frac{f}{1-f}; \quad f = (1-a)\left(\frac{D}{D+ks}\right)^2 \tag{6.6}$$

with the induction factor $a = 1 - u_h/u_{h0}$ and the mean turbine distance s. Such approaches decisively depend on the geometry of the wind parks and the wind direction relative to the orientation of the turbine rows. We do not want here to deal with the complications for special arrangements of turbines in a wind park, but we want to analyse the overall efficiency of very large wind parks. Therefore, we present an analytical top-down approach in Sect. 6.2 to derive the mean features dominating the efficiency of large wind parks.

Elliot and Barnard (1990), e.g., collected wind data at nine meteorological towers at the Goodnoe Hills MOD-2 wind turbine site to characterize the wind flow over the site both in the absence and presence of wind turbine wakes. The wind turbine wake characteristics analyzed included the average velocity deficits, wake turbulence, wake width, wake trajectory, vertical profile of the wake, and the stratification of wake properties as a function of the ambient wind speed and turbulence intensity. The wind turbine rotor disk at that site spanned a height of 15–107 m. The nine towers' data permitted a detailed analysis of the wake behaviour at a height of 32 m at various downwind distances from 2 to 10 rotor diameters (D). The relationship between velocity deficit and downwind distance was surprisingly linear [i.e. $n = 1$ in (6.1)], with average maximum deficits ranging from 34 % at 2 D to 7 % at 10 D. Largest deficits were at low wind speeds and low turbulence intensities. Average wake widths were 2.8 D at a downwind distance of 10 D. Implications for turbine spacing are that, for a wind park with a 10-D row separation, park efficiency losses would be significantly greater for a 2-D than a 3-D spacing because of incremental effects caused by overlapping wakes. Other interesting wake properties observed were the wake turbulence (which was greatest along the flanks of the wake). The vertical variation of deficits (which were greater below hub height than above), and the trajectory of the wake (which was essentially straight).

6.2 Analytical Model for Mean Wind Speed in Wind Parks

In the 1990s reasoning on nearly infinitely large wind parks was a purely academic exercise. Now, with the planning of large offshore wind farms off the coasts of the continents and larger islands such exercises have got much more importance (Barthelmie et al. 2005; Frandsen et al. 2006, 2009). In principle, two different approaches for modelling the effects of large wind parks are possible: a bottom-up approach and a top-down approach. The bottom-up approach is based on a superposition of the different wakes of the turbines in a wind park. It requires a good representation of each single wake (see Sect. 6.1) in a three-dimensional flow model (Lissaman 1979; Jensen 1983) and a wake combination model. Reviews are given in Crespo et al (1999) and Vermeer et al (2003). Numerically, this approach is supported by large-eddy simulations (LES) today (Wussow et al. 2007; Jimenez et al. 2007; Steinfeld et al. 2010; Troldborg et al. 2010).

The top-down approach considers the wind park as a whole as an additional surface roughness, as an additional momentum sink or as a gravity wave generator in association with a temperature inversion aloft at the top of the boundary layer (for the latter idea see Smith 2010), which modifies the mean flow above it (Newman 1977; Bossanyi et al. 1980; Frandsen 1992). Crespo et al. (1999) rates this latter class of models—although they have not been much used so far at that time—as being interesting for the prediction of the overall effects of large wind farms. Many of these models still have analytical solutions which make them attractive, although they necessarily contain considerable simplifications. Nevertheless, they can be used for first order approximations in wind park design. More detailed analyses require the operation of complex three-dimensional numerical flow models on large computers in the bottom-up approach.

Smith (2010) uses an analogy to atmospheric flow over a mountain range in order to derive his considerations. His model includes pressure gradients and gravity wave generation associated with a temperature inversion at the top of the boundary layer and the normal stable tropospheric lapse rate aloft. The pattern of wind disturbance is computed using a Fast Fourier Transform. The slowing of the winds by turbine drag and the resulting loss of wind farm efficiency is controlled by two factors. First is the size of the wind farm in relation to the restoring effect of friction at the top and bottom of the boundary layer. Second is the role of static stability and gravity waves in the atmosphere above the boundary layer. The effect of the pressure perturbation is to decelerate the wind upstream and to prevent further deceleration over the wind farm with a favourable pressure gradient. As a result, the wind speed reduction in Smith's (2010) approach is approximately uniform over the wind farm. In spite of the uniform wind over the farm, the average wind reduction is still very sensitive to the farm aspect ratio. In the special case of weak stability aloft, weak friction and the Froude Number close to unity, the wind speed near the farm can suddenly decrease; a phenomenon that Smith (2010) calls 'choking'. We will not follow this idea here. Rather, a top-down approach based on momentum extraction from the flow will be presented in more detail in this subchapter.

The derivation of the analytical wind park model shown here is an extension of earlier versions of this model documented in Frandsen (1992), Emeis and Frandsen (1993) and Emeis (2010a). The consideration of a simple, analytically solvable momentum balance of large wind parks in this subchapter will show that the design of a wind park and distance among each other has to take into account the properties of the surface on which they are erected and the thermal stability of the atmosphere typical for the chosen site. The momentum balance presented here will indicate that the distance between turbines in an offshore wind park and the distance between entire offshore parks must be considerably larger than for onshore parks. Turbines will be characterized only by their hub height, rotor diameter and thrust coefficient. Near wake properties are disregarded.

Starting point for the analytical wind park model is the overall mass-specific momentum consumption m of the turbines which is proportional to the drag of the turbines c_t and the wind speed u_h at hub height h:

$$m = c_t u_h^2 \tag{6.7}$$

In an indefinitely large wind park, this momentum loss can only be accomplished by a turbulent momentum flux τ from above. Here, u_0 is the undisturbed wind speed above the wind park, K_m is the momentum exchange coefficient and Δz is the height difference between hub height of the turbines and the undisturbed flow above the wind park (see Fig. 6.1):

$$\frac{\tau}{\rho} = K_m \frac{u_0 - u_h}{\Delta z} \tag{6.8}$$

The turbulent exchange coefficient K_m describes the ability of the atmosphere to transfer momentum vertically by turbulent motion. This coefficient describes an atmospheric conductivity giving the mass-specific momentum flux (physical units: m^2/s^2) per vertical momentum gradient (unit: 1/s). Thus K_m has the dimension of a viscosity (unit: m^2/s). Typical values of this viscosity are between 1 and 100 m^2/s. The main task in the formulation of the analytical park model is to describe the exchange coefficient K_m as function of the outer (surface roughness, thermal stratification of the boundary layer) and inner (drag of the turbines, turbulence generation of the turbines) conditions in the wind park. A major variable in this context is turbulence intensity T_i [see (3.10] for a definition) which is directly proportional to K_m. We obtain from the stability-dependent formulation of Monin–Obukhov similarity in the surface layer (see Sect. 3.1.1):

$$K_m = \kappa u_* z \frac{1}{\phi_m} \tag{6.9}$$

with the von Kármán constant $\kappa = 0.4$, the friction velocity u_* [see (A.13) in the Appendix], the height z and the stability function ϕ_m:

$$\frac{1}{x} \quad \text{for} \frac{z}{L_*} < 0$$

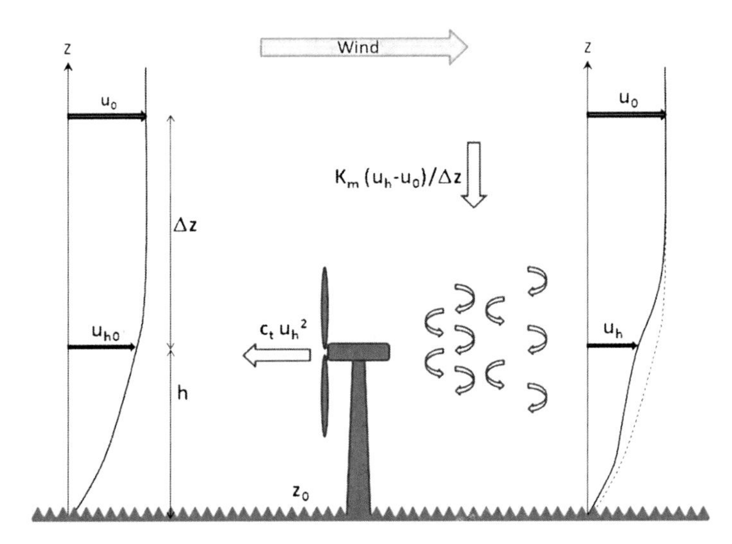

Fig. 6.1 Schematic of momentum loss and replenishment in an indefinitely large wind park. The undisturbed flow is approaching from *left*

$$\phi_m\left(\frac{z}{L_*}\right) = 1 \quad \text{for}\, \frac{z}{L_*} = 0 \tag{6.10}$$

$$1 + a\frac{z}{L_*} \quad \text{for}\, \frac{z}{L_*} > 0$$

with $x = (1 - b\ z/L_*)^{1/4}$ and the Obukhov length L_* defined in (3.11). We use $a = 5$ and $b = 16$ in (6.10). Assuming a logarithmic wind profile, the friction velocity, u_* is given by:

$$u_* = u_h \kappa \left(\ln\left(\frac{h}{z_0}\right) - \Psi\left(\frac{\mathrm{h}}{\mathrm{L}_*}\right)\right)^{-1} \tag{6.11}$$

where Ψ is given by (3.15) for unstable conditions and the first equation of (3.21) for neutral and stable conditions.

Following Frandsen (2007), we define the wind park drag coefficient, c_t as a function of the park area A, the rotor area $0.25\pi D^2$, the number of turbines N and the turbine thrust coefficient C_T:

$$c_t = \frac{1}{8}\frac{N\pi D^2}{A} C_T \tag{6.12}$$

C_T is about 0.85 for lower wind speeds around the cut-in wind speed and decreases around and above the rated wind speed of the turbines with increasing wind speed (Barthelmie et al 2006; Jimenez et al 2007). The exact value depends on the construction of the turbine and its operation. We use the following empirical

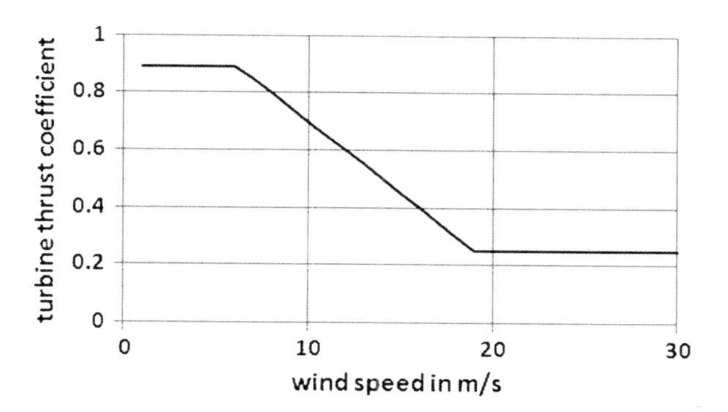

Fig. 6.2 Wind-speed dependent turbine thrust coefficient [see Eq. (6.13)] used in the simple analytical model park model

relation for the thrust coefficient (taken from Fig. 9 in Magnusson 1999) and additionally consider the maximal value at Betz's limit[1]:

$$C_T = \min(\max(0.25; 0.5 + 0.05(14 - u_h)); 0.89) \tag{6.13}$$

Due to (6.13), c_t depends on u_h (see Fig. 6.2) and we have to iterate at least once when we want to solve for u_h later.

The reduction of wind speed in hub height h in the park interior does not only depend on the turbine drag coefficient c_t but also on the roughness of the surface underneath the turbines. This surface roughness can be described by a surface drag coefficient, $c_{s,h}$ observed at height h by rearranging (6.11):

$$c_{s,h} = u_*^2/u_h^2 = \kappa^2\left(\ln\left(\frac{h}{z_0}\right) - \Psi\left(\frac{\mathrm{h}}{\mathrm{L}_*}\right)\right)^{-2} \tag{6.14}$$

Turbine drag and surface drag can be combined in an effective drag coefficient:

$$c_{teff} = c_t + c_{s,h}. \tag{6.15}$$

There are two ratios describing the wind reduction in the wind park. The reduction of the wind speed at hub height compared to the undisturbed wind speed aloft is denoted by R_u:

$$R_u = \frac{u_h}{u_0} \tag{6.16}$$

The reduction of the wind speed at hub height compared to the undisturbed wind speed upstream of the wind park in the same height h, u_{h0} is denoted by R_t:

[1] The thrust coefficient is the ratio of resistance force T to the dynamic force $0.5\rho u^2{}_0 D$ (rotor area D). The resistance force of an ideal turbine is given by $T = 0.5\rho u^2{}_0 A[4r(1-r)]$ with $r = (u_0 - u^*_h)/u_0$. u^*_h is the mean of u_h and u_0. We have $u^*_h = u_0\,(1-r)$. Thus, $C_T = [4r(1-r)]$. For $u_h = 0$ it follows $u^*_h = 0.5u_0$, $r = 0.5$ and $C_T = 1$. For $u_h = u_0$ follows $u^*_h = u_0$, $r = 0$ and $C_T = 0$. The yield is $P = Tu^*_h\,0.5 = \rho u_0^3 A[4r(1-r)^2]$ and the yield coefficient is $C_P = [4r(1-r)^2]$. For optimal yield at the Betz's limit is r = 1/3 (calculated from $\partial C_P(r)/\partial r = 0$) and $C_T = 8/9$ (Manwell et al. 2009)

$$R_t = \frac{R_u(c_{teff})}{R_u(c_{s,h})} \tag{6.17}$$

using $R_u(c_{s,h}) = u_{h0}/u_0$. Inserting for the exchange coefficient K_m (6.9) and the effective drag coefficient (6.15) in (6.7) yields:

$$c_{teff} u_h^2 = \frac{\kappa u_* z(u_0 - u_h)}{\Delta z \phi_m} \tag{6.18}$$

The height z in (6.18) is essentially $h + \Delta z$, so that the ratio $z/\Delta z$ can be approximated by a constant value:

$$\frac{z}{\Delta z} = f_{h,\Delta z} \tag{6.19}$$

The horizontal turbulence intensity I_u at hub height h is defined by:

$$I_u = \frac{\sigma_u}{u_h} \tag{6.20}$$

The standard deviation of the horizontal wind speed can be parameterized using the friction velocity u_*:

$$\sigma_u = \frac{1}{\kappa} u_* \tag{6.21}$$

which yields the following relation between friction velocity, u_* and turbulence intensity, I_u:

$$u_* = \kappa \sigma_u = \kappa u_h I_u \tag{6.22}$$

Inserting of (6.19) and (6.22) in (6.18) yields finally:

$$c_{teff} u_h^2 = \frac{\kappa^2 u_h (u_0 - u_h)}{\phi_m} f_{h,\Delta z} I_u = \frac{\kappa^2 u_h u_0^2}{u_0 \phi_m} f_{h,\Delta z} I_u - \frac{\kappa^2 u_h^2}{\phi_m} f_{h,\Delta z} I_u \tag{6.23}$$

Rearrangement leads to:

$$c_{teff} u_h^2 + \frac{\kappa^2 u_h^2}{\phi_m} f_{h,\Delta z} I_u = u_h^2 \left(c_{teff} + \frac{\kappa^2}{\phi_m} f_{h,\Delta z} I_u \right) = u_0^2 \frac{R_u \kappa^2}{\phi_m} f_{h,\Delta z} I_u \tag{6.24}$$

and finally to an expression for the ratio (6.16):

$$R_u = \frac{u_h}{u_0} = \frac{u_h^2}{R_u u_0^2} = \frac{\frac{\kappa^2}{\phi_m} f_{h,\Delta z} I_u}{\left(c_{teff} + \frac{\kappa^2}{\phi_m} f_{h,\Delta z} I_u \right)} = \frac{f_{h,\Delta z} I_u}{\left(f_{h,\Delta z} I_u + \frac{\phi_m}{\kappa^2} c_{teff} \right)} \tag{6.25}$$

Thus, the ratio (6.17) between the wind speed at hub height inside the wind park to the undisturbed wind speed upstream is:

$$R_t = \frac{\left(f_{h,\Delta z} I_u + \frac{\phi_m}{\kappa^2} c_{s,h}\right)}{\left(f_{h,\Delta z} I_u + \frac{\phi_m}{\kappa^2} c_{teff}\right)} \tag{6.26}$$

Formulation (6.26) permits easily to add the turbulence intensity produced by the turbines during operation to the upstream turbulence intensity ($I_{u,eff}^2 = I_{u0}^2 + I_{u,t}^2$). Following Barthelmie et al. (2003) the additional turbulence, $I_{u,t}$ can be parameterized as a function of the thrust coefficient (6.13) using a mean turbine distance normalized by the turbine diameter s:

$$I_{u,t} = \sqrt{\frac{1,2C_T}{s^2}} \tag{6.27}$$

The upper frame in Fig. 6.3—in displaying R_t from (6.24)—shows how much the wind speed at hub height will be reduced as a function of the atmospheric instability and the surface roughness. The presented results have been found for turbines with a hub height of 92 m, a rotor diameter of 90 m and a mean distance between two turbines in the park of 10 rotor diameters. It becomes obvious that the reduction is smallest (a few percent) for unstable thermal stratification of the atmospheric boundary layer and high surface roughness. I.e., the reduction is smallest over a rough land surface with trees and other obstacles for cold air flowing over a warm surface (usually during daytime with strong solar insolation). The largest reduction (up to 45 %) occurs for very smooth sea surfaces when warm air flows over cold waters. This may happen most preferably in springtime. The lower frame of Fig. 6.3 translates this wind speed reduction into a reduction of the available wind power by plotting the third power of R_t from (6.26). The strong stability dependence of the reduction of the available power can be confirmed from measurements at the Nysted wind park in Denmark (Barthelmie et al. 2007).

The dependence of wind and available power reduction as function of surface roughness has consequences for offshore wind parks which will become the major facilities for wind power generation in the near future. The lower turbulence production due to the relative smoothness of the sea surface compared to land surfaces hampers the momentum re-supply from the undisturbed flow above. In order to limit the wind speed reduction at hub height in the interior of the wind park to values known from onshore parks, the turbines within an offshore wind park must have a larger spacing than within an onshore park. Roughly speaking, the number of turbines per unit area in an offshore park with roughness $z_0 = 0.001$ m must be approximately 40 % lower than in an onshore park with $z_0 = 0.1$ m in order to have the same power yield for a given wind speed and atmospheric stability.

Inversely, Eq. (6.26) may be used to determine the optimal areal density of turbines in a large wind park for given surface roughness and atmospheric stability conditions.

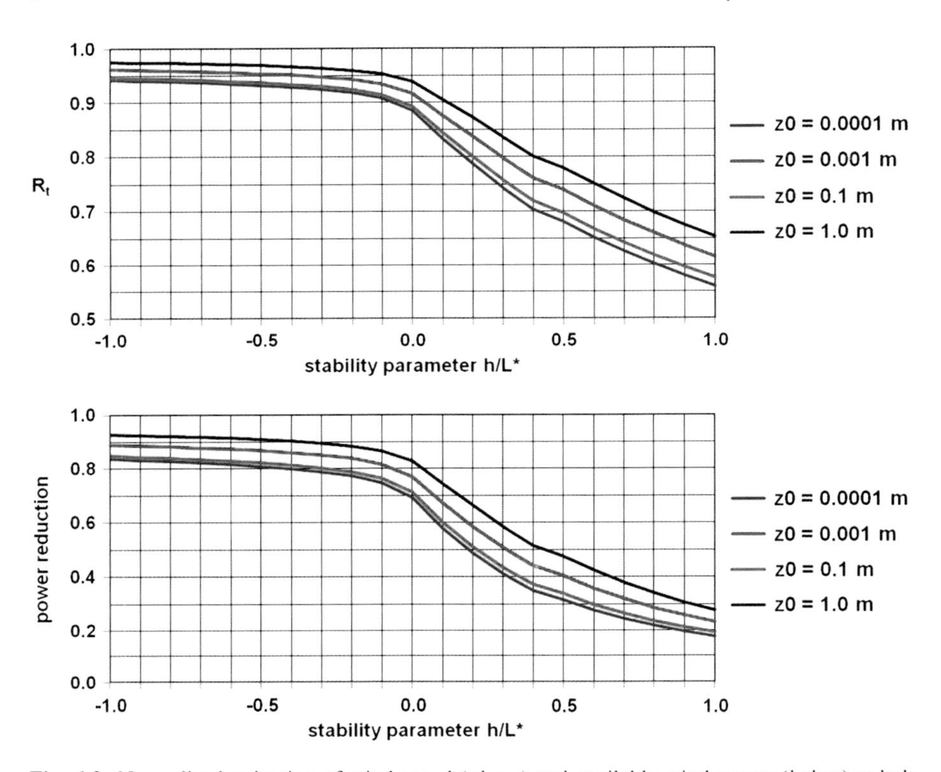

Fig. 6.3 Normalised reduction of wind speed (*above*) and available wind power (*below*) at hub height in an indefinitely large wind park as function of atmospheric instability (h/L* = 1: strong instability, 0: neutral stability, + 1: stable stratification) and surface roughness (z0 = 0.0001 m: very smooth sea surface, 0.001 m: rough sea surface, 0.1 m: smooth land surface, 1.0 m: rough land surface)

6.3 Analytical Model for Wind Park Wakes

The estimation of the length of the wakes of large wind parks is essential for the planning of the necessary distance between adjacent wind parks. This estimation can be made using the same principal idea as in the subchapter before: the missing momentum in the wake of an indefinitely broad wind park can only be replenished from above (Fig. 6.4). If we imagine to move with an air parcel, then we feel the acceleration of the speed of this parcel, u_{hn} from u_{hn0} at the rear end of the park to the original undisturbed value, u_{h0}, which had prevailed upstream of the park (neglecting the Coriolis force):

$$\frac{\partial u_{hn}}{\partial t} = \frac{\partial(\tau/\rho)}{\partial z} \tag{6.28}$$

Substituting the differentials by finite differences and using (6.8) leads to:

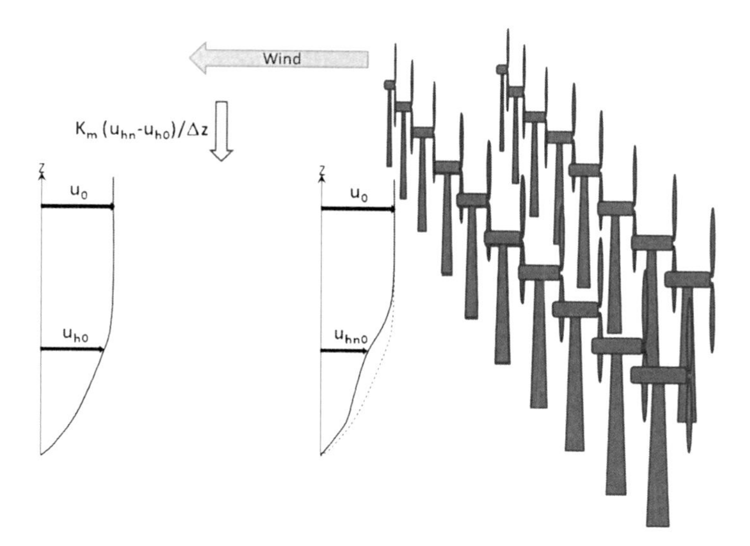

Fig. 6.4 Wind speed u_{hn} (from u_{hno} at the rear end of the wind park to the original undisturbed wind speed further down to the *left* u_{h0}) in the wake of an indefinitely broad wind park

$$\frac{\Delta u_{hn}}{\Delta t} = \frac{\kappa u_* z}{\Delta z^2}(u_{h0} - u_{hn}) = \frac{\kappa u_* z u_{h0}}{\Delta z^2} - \frac{\kappa u_* z u_{hn}}{\Delta z^2} \tag{6.29}$$

This is a first order difference equation of the form:

$$\frac{\Delta u_{hn}}{\Delta t} + \alpha u_{hn} = \alpha u_{h0} \tag{6.30}$$

with $\alpha = \kappa u_* z/\Delta z^2$ and the time-dependent solution:

$$u_{hn}(t) = u_{h0} + C\exp(-\alpha(t - t_0)) \tag{6.31}$$

The constant of integration C can be determined from the initial condition:

$$u_{hn}(t = t_0) = u_{hn0} = u_{h0} + C \tag{6.32}$$

Please note the difference between the undisturbed wind speed u_{h0} at hub height and the wind speed at hub height directly behind the wind park u_{hn0}. Inserting (6.32) in (6.31) yields:

$$u_{hn}(t) = u_{h0} + (u_{hn0} - u_{h0})\exp(-\alpha t) \tag{6.33}$$

Dividing by u_{h0} gives the ratio R_n between the wind speed at hub height in the wake to the undisturbed wind speed in the same height u_{h0}:

$$R_n = \frac{u_{hn}(t)}{u_{h0}} = 1 + \left(\frac{u_{hn0}}{u_{h0}} - 1\right)\exp(-\alpha t) \tag{6.34}$$

The factor α in (6.30), (6.33) and (6.34) depends on the surface roughness and the thermal stratification of the boundary layer via (6.11). This solution is in the time domain. It can be converted into the space domain by assuming an average wind speed over the wake.

The upper frame in Fig. 6.5 shows wake lengths as function of surface roughness for neutral stability ($h/L^* = 0$) by plotting the third power of R_n from (6.34). If we define the distance necessary for a recovery of the available power to 95 % of its undisturbed value upstream of the park as wake length, then we see a wake length of 4 km for rough land surfaces and a wake length of about 18 km for smooth sea surfaces. Figure 6.5 has been produced for the same park parameters as Fig. 6.3. Actually, the results from Fig. 6.3 serve as left boundary conditions for Fig. 6.5. The lower frame of Fig. 6.5 demonstrates the strong influence of atmospheric stability on the wake length for an offshore wind park over a smooth sea surface ($z_0 = 0.0001$ m). Taking once again the 95 % criterion, the wake length for very unstable atmospheric conditions is still about 10 km. For very stable conditions, the wake length is even longer than 30 km. Such long wakes have been confirmed from satellite observations (Christiansen and Hasager 2005).

6.4 Application of the Analytical Model with FINO1 Stability Data

The application of the above analytical model to a real wind park needs the knowledge of the frequency distribution of atmospheric stabilities at the site of the wind park. We give here an example by using the distribution measured at 80 m height at the mast FINO1 in the German Bight for the years 2005 and 2006. Figure 6.6 shows this distribution for the range $-2 \leq z/L_* \leq 2$. 91.16 % of all data fall into this range. The highest frequency occurs for the bin $-0.15 \leq z/L_* \leq -0.05$. The median of the full distribution is at $z/L_* = -0.11$, the median of the range shown in Fig. 6.6 is $z/L_* = -0.07$. Now the above equations for the reduction of wind speed in the park interior (6.26) and the wake length (6.34) are solved for all 41 bins shown in Fig. 6.6 and the resulting values for R_t and R_n are multiplied with the respective frequencies from Fig. 6.6.

Rebinning the resulting R_t and R_n values leads to the distributions shown in Figs. 6.7 and 6.8. The top frame in Fig. 6.7 shows the distribution of wind speed reductions at hub height in the park interior. The most frequent speed reduction R_t is 0.95, the median is 0.93 and the weighted mean is 0.87. The 90th percentile is observed at 0.73 and the 95th percentile at 0.65. The lower frame of Fig. 6.7 gives the resulting reductions in power yield. The most frequent power yield reduction is 0.83, the median is 0.80 and the weighted mean is 0.70. The 90th percentile is observed at 0.37 and the 95th percentile at 0.24.

Figure 6.8 displays the respective distribution of the wake length. Here, the wake length has been defined as above as the distance where the power yields have

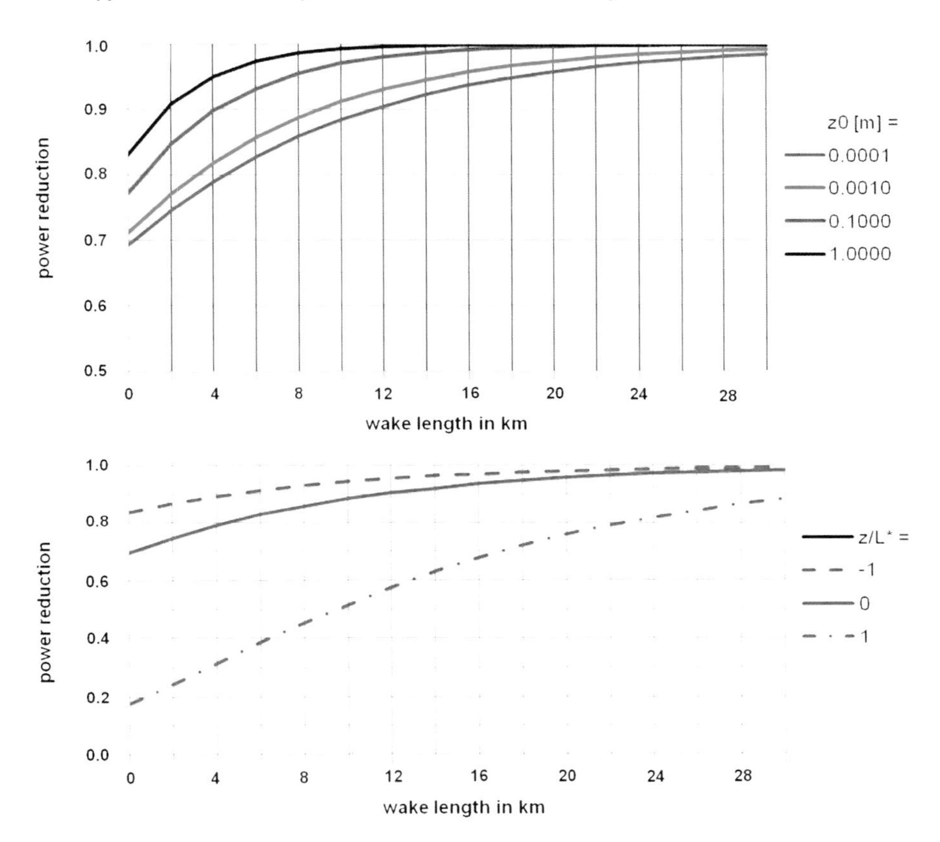

Fig. 6.5 Normalised reduction of available wind power at hub height behind an indefinitely large wind park as function of the distance from the rear side of the park. *Above*: as function of surface roughness (z0 = 0.0001 m: very smooth sea surface, 0.001 m: rough sea surface, 0.1 m: smooth land surface, 1.0 m: rough land surface) with neutral stability. *Below*: as function of atmospheric instability ($h/L^* = -1$: strong instability, 0: neutral stability, + 1: stable stratification) for a smooth sea surface

recovered to 95 % of their original value upstream of the park. The most frequent wake length is 11 km, the median is nearly 14 km and the weighted mean is 17.7 km. The 90th percentile is observed at 31 km and the 95th percentile at 37 km.

6.5 Risks that a Tornado Hits a Wind Park

Tornadoes are a risk for wind turbines. The weakest (F1) tornadoes have a wind speed of 32–50 m/s, while F2 tornadoes reach 70 m/s which is well above the survival speed of wind turbines. But even if the peak wind speed is below the

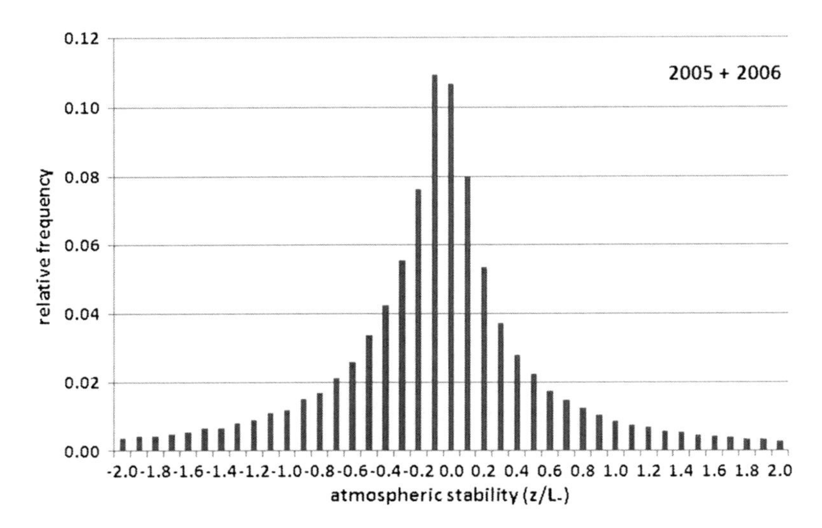

Fig. 6.6 Frequency distribution of atmospheric stability at 80 m height at the mast FINO1 in the German Bight. Bin width is 0.1

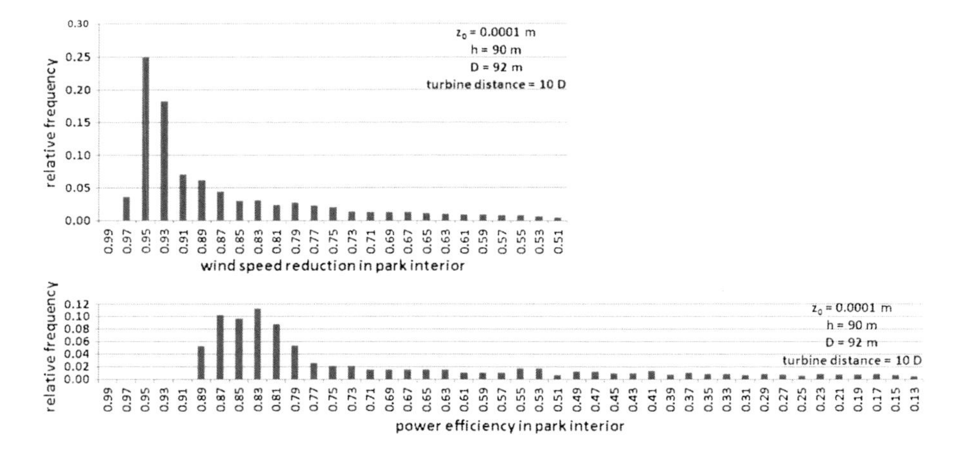

Fig. 6.7 Frequency distribution of wind speed reduction at hub height in the park interior (*top*) and of power yield reduction (*below*) using the stability data from Fig. 6.6. Bin width is 0.02

survival speed, the most dangerous feature is the rapid increase of wind speed connected with a rapid wind direction change when a tornado approaches a wind turbine. There is no reasonable alert time available.

Dotzek et al (2010) investigate the risk that an offshore wind park in the German Bight will be hit by a waterspout. Assuming an area of about 100 km^2 ($10 \times 10\ km^2$) as typical for prospective offshore wind parks off the German

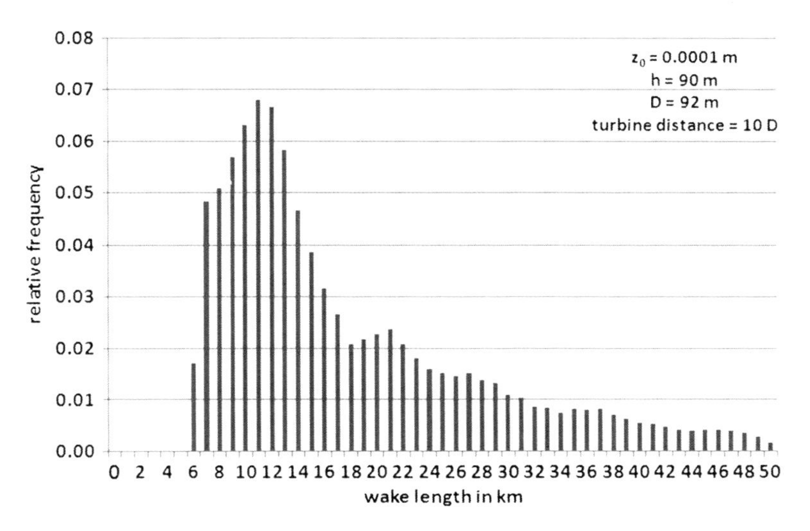

Fig. 6.8 Frequency distribution of the wake length of an indefinitely broad wind park in km using the stability data from Fig. 6.6. Bin width is 1 km. Wake length has been defined as the distance where 95 % of the original power yield is reached again

coast, the probability is estimated that such a wind park will be affected by waterspouts. This estimation does not look for the probability that a single wind turbine is hit by the vortex centre, i.e. the probability of a mathematical point being hit (Thom 1963) is not investigated. Due to the horizontal wind shear across the vortex' core and mantle regions, even a near miss by a waterspout may be hazardous for a wind turbine. In addition, it is presently unclear if the small-scale wind field in a wind park altered by the wind turbine wakes themselves (Christiansen and Hasager 2005) may actually increase the likelihood of a hit once a waterspout enters an array of wind turbines. Therefore, the recurrence time of a waterspout anywhere within the wind park instead of at an individual wind turbine site is analysed.

Taking the waterspout incidence presently known for the German North Sea coast [which is about one tornado per 10,000 km^2 per year, based on estimates by Koschmieder (1946) or Dotzek (2003)], one can expect one tornado in an offshore wind park once within one hundred years. This includes the assumption that waterspouts occur homogeneously over the German Bight area. If using the upper limit of Koschmieder's estimate, which is two waterspouts per 10,000 km^2 per year, this recurrence time reduces to 50 years for a single wind park.

While this still seems to be a long interval, one has to take into account that the total area of approved or actual off-shore wind parks in the German Bight is already 648 km^2 in 2010 (Source: German Federal Maritime and Hydrographic Agency; Bundesamt für Seeschifffahrt und Hydrographie), leading to a recurrence interval of less than eight years for any wind park to be hit by waterspouts in a given year, based on Koschmieder's incidence estimate of two waterspouts per

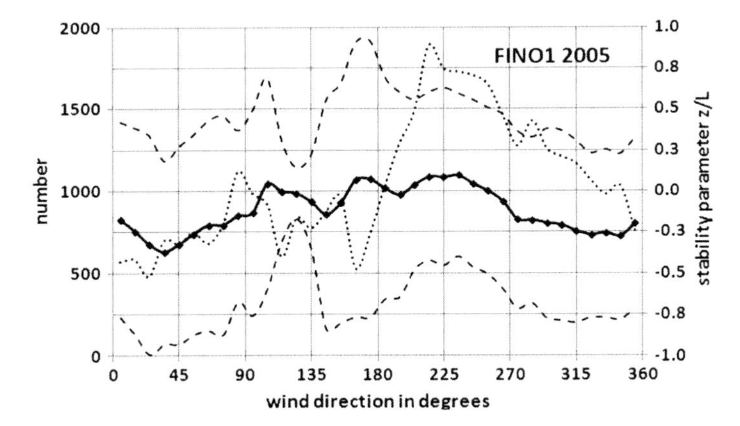

Fig. 6.9 Example for the dependence of the mean stability of the marine boundary layer air for different wind directions from FINO1 data for the year 2005. The full line gives the annual mean stability parameter h/L^* (*right-hand axis*), the dashed lines give the annual mean minus and plus one standard deviation of this stability parameter. The dotted line gives the number of 10 min data per 10 degree wind direction interval (*left-hand axis*)

year per 10,000 km^2. A recent report by the European Wind Energy Association (EWEA 2007) identified that offshore North Sea wind parks with an area of 17,900 km^2 were needed to supply 180 GW, i.e. about 25 % of Europe's current electricity needs. A scenario for 2020 foresees the installation of 40 GW, which would require about 3,980 km^2 of wind parks. Should this scenario materialise, one or more waterspouts within an offshore wind park would have to be expected every other year.

6.6 Summary for Wind Parks

The roughness of the underlying surface on which large wind parks are erected turns out to be a decisive parameter governing the efficiency of such parks. This happens, because the ability of the atmosphere to supply momentum from the undisturbed flow above depends on turbulence intensity, which increases with increasing surface roughness. Therefore, in offshore wind parks, this supply is much less than over land, where turbulence intensity is much higher. Thus, in offshore wind parks, the spacing between the turbines in the park must be larger as onshore. The gaps between adjacent offshore wind parks must be larger as well.

Another important governing parameter for the efficiency of wind parks is the thermal stability of the atmosphere, because turbulence intensity is much higher for unstable stratification than for stable stratification. Over the ocean stability mainly depends on the type of thermal advection. Cold air advection over warmer water usually leads to unstably stratified boundary layers and warm air advection over cold water to stably stratified boundary layers. In the west wind belts of the temperate latitudes, cold and warm air advection regimes are coupled to different wind directions which correspond to the typical wind directions in the warm and cold sectors of the moving depressions (see Fig. 6.9 for an example). As mean turbine distances and gaps between entire parks can be smaller for unstable stratification than for stable stratification, it might be advisable to make at least the gaps between entire offshore parks wind direction-dependent having larger gaps in the direction of flow that is connected to warm air advection.

The example in Fig. 6.9 shows unstable stratification for north-westerly and northerly winds and stable stratification for south-westerly winds. In such a wind regime, it might be advisable to have larger distances between the turbines and between wind parks in the south-west to north-east direction, while shorter distances are possible in the north-west to south-east direction. The lower frame of Fig. 6.3 shows that there is a factor of two in power reduction between $h/L_* = -0.3$ and $h/L_* = 0.1$, which are the typical mean stabilities in Fig. 6.9. Therefore, the analysis of the relation between average stability of the boundary layer and the wind direction should be analysed during the siting procedure for offshore wind parks. This advice does not apply to onshore wind parks, because here the atmospheric stability mainly depends on cloudiness and time of the day, but not so much on wind direction.

References

Barthelmie, R.J., L. Folkerts, F.T. Ormel, P. Sanderhoff, P.J. Eecen, O. Stobbe, N.M. Nielsen: Offshore Wind Turbine Wakes Measured by Sodar. J. Atmos. Oceanogr. Technol. **20**, 466–477 (2003)

Barthelmie, R., Frandsen, S.T., Rethore, P.E., Jensen, L.: Analysis of atmospheric impacts on the development of wind turbine wakes at the Nysted wind farm. Proc. Eur. Offshore Wind Conf. 2007, Berlin 4.-6.12.2007 (2007)

Barthelmie, R., Hansen O.F., Enevoldsen K., Højstrup J., Frandsen S., Pryor S., Larsen S.E., Motta M., and Sanderhoff P.: Ten Years of Meteorological Measurements for Offshore Wind Farms. J. Sol. Energy Eng. 127, 170–176 (2005)

Barthelmie R.J., L.E. Jensen: Evaluation of wind farm efficiency and wind turbine wakes at the Nysted offshore wind farm. Wind Energy **13**, 573–586 (2010)

Barthelmie, R.J., S. Pryor, S. Frandsen, S. Larsen: Analytical Modelling of Large Wind Farm Clusters. Poster, Proc. EAWE 2004 Delft (2004). (http://www.risoe.dk/vea/storpark/Papers%20and%20posters/delft_013.pdf)

Bossanyi, E.A., Maclean C., Whittle G.E., Dunn P.D., Lipman N.H., Musgrove P.J.: The Efficiency of Wind Turbine Clusters. Proc. Third Intern. Symp. Wind Energy Systems, Lyngby (DK), August 26–29, 1980, 401–416 (1980)

Christiansen, M.B., Hasager, C.B.: Wake effects of large offshore wind farms identified from satellite SAR. Rem. Sens. Environ. 98, 251–268 (2005)

Crespo, A., Hernandez, J., Frandsen, S.: Survey of Modelling Methods for Wind Turbine Wakes and Wind Farms. Wind Energy **2**, 1–24 (1999)

Dotzek, N., S. Emeis, C. Lefebvre, J. Gerpott: Waterspouts over the North and Baltic Seas: Observations and climatology, prediction and reporting. Meteorol. Z. **19**, 115–129 (2010)

Dotzek, N.: An updated estimate of tornado occurrence in Europe. – Atmos. Res. **67–68**, 153–161 (2003)

Elliot, D.L., J.C. Barnard: Observations of Wind Turbine Wakes and Surface Roughness Effects on Wind Flow Variability. Solar Energy, **45**, 265–283(1990)

Elliot, D.L.: Status of wake and array loss research. Report PNL-SA–19978, Pacific Northwest Laboratory, September 1991, 17 pp. (1991) (available from: http://www.osti.gov/energycitations/product.biblio.jsp?osti_id=6211976)

Emeis, S., S. Frandsen: Reduction of Horizontal Wind Speed in a Boundary Layer with Obstacles. Bound.-Lay. Meteorol. **64**, 297–305 (1993)

Emeis, S.: A simple analytical wind park model considering atmospheric stability. Wind Energy **13**, 459–469 (2010a)

EWEA (Eds.): Delivering Offshore Wind Power in Europe. – Report, European Wind Energy Association, Brussels, 32 pp. (2007) [Available at www.ewea.org/fileadmin/ewea_documents/images/publications/offshore_report/ewea-offshore_report.pdf]

Frandsen, S., Jørgensen, H.E., Barthelmie, R., Rathmann, O., Badger, J., Hansen, K., Ott, S., Rethore, P.E., Larsen, S.E., Jensen, L.E.: The making of a second-generation wind farm efficiency model-complex. Wind Energy **12**, 431–444 (2009)

Frandsen, S.: On the Wind Speed Reduction in the Center of Large Cluster of Wind Turbines. J. Wind Eng. Ind. Aerodyn. **39**, 251–265 (1992)

Frandsen, S.: Turbulence and turbulence generated structural loading in wind turbine clusters. Risø-R-1188(EN), 130 pp. (2007)

Frandsen, S.T., Barthelmie, R.J., Pryor, S.C., Rathmann, O., Larsen, S., Højstrup, J., Thøgersen, M.: Analytical modelling of wind speed deficit in large offshore wind farms. Wind Energy **9**, 39–53 (2006)

Jensen, N.O.: A Note on Wind Generator Interaction. Risø-M-2411, Risø Natl. Lab., Roskilde (DK), 16 pp. (1983) (Available from http://www.risoe.dk/rispubl/VEA/veapdf/ris-m-2411.pdf

Jimenez, A., A. Crespo, E. Migoya, J. Garcia: Advances in large-eddy simulation of a wind turbine wake. J. Phys. Conf. Ser., **75**, 012041. DOI: 10.1088/1742-6596/75/1/012041(2007)

Koschmieder, H.: Über Böen und Tromben (On straight-line winds and tornadoes). Die Naturwiss. **34**, 203–211, 235–238 (1946) [In German]

Lissaman, P.B.S.: Energy Effectiveness of arbitrary arrays of wind turbines. AIAA paper 79-0114 (1979)

Magnusson, M.: Near-wake behaviour of wind turbines. J. Wind Eng. Ind. Aerodyn. **80**, 147–167 (1999)

Manwell, J.F., J.G. McGowan, A.L. Rogers: Wind Energy Explained: Theory, Design and Application. 2nd edition. John Wiley & Sons, Chichester. 689 pp. (2010)

Newman, B.G.: The spacing of wind turbines in large arrays. J. Energy Conversion **16**, 169–171 (1977)

Quarton, D.C.: Characterization of wind turbine wake turbulence and its implications on wind farm spacing. Final Report ETSU WN 5096, Department of Energy of the UK. Garrad-Hassan Contract (1989)

Smith, R.B.: Gravity wave effects on wind farm efficiency. Wind Energy, **13**, 449–458 (2010).

Steinfeld, G., Tambke, J., Peinke, J., Heinemann, D.: Application of a large-eddy simulation model to the analysis of flow conditions in offshore wind farms. Geophys. Res. Abstr. **12**, EGU2010-8320 (2010)

Thom, H.C.S.: Tornado probabilities. – Mon. Wea. Rev. **91**, 730–736 (1963)

Troen, I., E.L. Petersen: European Wind Atlas. Risø National Laboratory, Roskilde, Denmark. 656 pp. (1989)

Troldborg, N., J.N. Sørensen, R. Mikkelsen: Numerical simulations of wake characteristics of a wind turbine in uniform inflow. Wind Energy **13**, 86–99 (2010)

Vermeer, L.J., J.N. Sørensen, A. Crespo: Wind turbine wake aerodynamics. Progr. Aerospace Sci. **39**, 467–510 (2003)

Wussow, S., L. Sitzki, T. Hahm: 3D-simulations of the turbulent wake behind a wind turbine. J. Phys. Conf. Ser., **75**, 012033, DOI: 10.1088/1742-6596/75/1/012033 (2007)

Chapter 7
Outlook

This chapter is not designed to summarize the main points from the preceding chapters. This has already been done in the concluding subchapters of each of the Chaps. 3–6. Rather we will try to look briefly at possible future developments and a few limitations for the use of the material in this book. This concerns technical aspects as well as assessment methods for meteorological conditions and possible climate impacts of large-scale wind energy conversion.

7.1 Size of Wind Turbines

The evolution of wind turbines addressed in the introduction has not yet come to a halt. Larger and larger turbines are being designed and erected (Thresheret al. 2007). Turbines are increasing in hub height as well as in rotor diameter. The former involves new concepts for turbine towers, the latter depends critically on the availability of suitable blades (Grujicic et al. 2010). This development is fostered by two aspects. One issue is that the deployment of offshore wind turbines is very expensive and complicated. The foundation of the turbine masts in the sea floor (see, e.g., Wichtmann et al. 2009) and the transport by large vessels are still challenging tasks which have not been solved finally so far (Bretton and Moe 2009). In order to limit deployment costs, fewer but larger turbines are erected offshore. The other issue is that turbines are being erected more and more in less favourable wind climates, because the best and windiest sites near the coast are already in use and because wind power is needed in urban and industrial centres far away from the coasts as well. In order to get the same harvest from the turbines as in coastal windy areas, they must have larger hub heights to reach atmospheric levels with sufficient wind speed for an economically meaningful operation. Both developments lead to an increasing importance of the exact specification of the meteorological conditions described in this publication for siting and operation of these turbines. Nearly all new turbines will operate in the Ekman layer of the atmospheric boundary layer. For example,

S. Emeis, *Wind Energy Meteorology*, Green Energy and Technology,
DOI: 10.1007/978-3-642-30523-8_7,

the influence of nocturnal low-level jets on the energy production from wind turbines will grow beyond that what is experienced today.

7.2 Size of Offshore Wind Parks

The growing energy demand of mankind together with the limited resources of fossil fuels, the decreasing availability of suitable onshore sites for wind energy conversion and the necessity to bundle power transportation lines from the wind parks to the shore will continuously foster the planning and erection of huge offshore wind parks. The United Kingdom and Germany have already presented initiatives to erect large offshore parks. Many other countries, especially those having ocean coastlines in temperate latitude will follow. The larger these wind parks become, the more the simple analytical estimations presented in Chap. 6 of this publication will become relevant. This is because the conditions in very large wind parks are much closer to the assumptions made for these analytical estimations than in the presently existing parks.

7.3 Other Techniques of Converting Wind Energy

The meteorological basics gathered in this publication are relevant for all boundary layer applications which depend on the kinetic energy contained in the winds. The presented wind and turbulence laws and distributions influence classical wind turbines (regardless whether they have a horizontal or a vertical rotor axis) as well as classical or new sailing boats and new kite-torn ships. However, applications relying on kites soaring several kilometres about the surface are beyond the scope of this publication. Existing climatologies of upper air winds above the atmospheric boundary layer have to be investigated for the planning and operation of the latter installations. These upper air winds are principally described by the laws for geostrophic, gradient and thermal winds given in Sects. 2.3 and 2.4.

7.4 New Measurement and Modelling Tools to Assess Wind Conditions

Measurement techniques for atmospheric parameters at hub height and over the area swept by the rotor must change in future. The growing hub heights and upper tip heights of the turbine rotors make it more and more impossible to perform in situ measurements from masts specially erected for this purpose. Ground-based remote sensing will substitute mast measurements in the foreseeable future. Emeis (2010b, 2011) gives an overview of the present abilities to probe the atmospheric boundary layer by ground-based remote sensing. The substitution process from in situ to remote sensing measurements is to be accompanied by scientific investigations which compare the wind and turbulence data obtained from masts and remote sensing techniques. Such investigations are presently under way and have to lead to rewritten standards for measurement procedures. Most probably optical techniques such as wind lidars will be the measurement tools for the future (see, e.g., Trujillo et al. 2011).

The abilities of numerical models must be enhanced as well. Simple analytical models such as those presented in this publication (see, e.g., Sect. 4.2) and existing mesoscale wind field models will no longer be sufficient for large turbines in very complex terrain and for turbines in smaller wind parks. Work is under way to design more sophisticated models which have a higher spatial resolution, both in the horizontal and in the vertical close to the ground. This work includes the development of suitable large-eddy simulation (LES) models for offshore wind parks (Cañadillas and Neumann 2010; Steinfeld et al. 2010) and for smaller wind parks and complex terrain.

7.5 Wind Resources and Climate Change

Wind turbines and wind parks are usually planned for several decades of operation. Thus, estimations on future changes in wind resources in selected regions may influence the economic prospects of these installations. Site assessment, especially for regions with marginal wind resources, should take into account future wind scenarios from global and regional climate models.

First of all, global warming is expected to generally weaken the west wind belts around the globe, because the warming in the polar regions will be stronger than in the tropics. This differential warming trend will decrease the global meridional temperature gradient between the lower and the higher latitudes, which had been identified as the main driver for the global westerlies in Sect. 2.1. Due to nonlinearities in the atmospheric system, this relation is not straight-forward and needs specific investigations (see, e.g., Geng and Sugi 2003). Additionally, the weakening temperature gradient could also be accompanied by a poleward shift of the climate zones and storm tracks on Earth (Yin 2005). These two effects can be derived from simulations with global climate models.

Apart from the general impact on the global meridional temperature gradients, climate change can also lead to regional atmospheric circulation changes. These changes may alter regional weather patterns such as regional storm tracks and main wind directions which can lead to considerable variations in the wind climate of a selected site. The assessment of such possible regional circulation changes should be made from regional climate model simulations. Regional climate models have a much higher spatial resolution than global climate models. Regional models are run for limited regions taking the output from global climate models as boundary conditions. Many of such regional studies have been performed. For a wind energy-related study, see, e.g., Nolan et al. (2011).

7.6 Repercussions of Large-Scale Wind Power Extraction on Weather and Climate

Large-scale exploitation of wind energy will probably have impacts on regional winds. Large wind farms increase the surface roughness and the surface drag and thus change the local and regional momentum budgets. This interaction has been shown in Chap. 6. More challenging is the investigation of global effects. If the extracted energy comes close to the level of the totally available wind energy (see Sects. 1.4 and 1.5 above), it will definitely have an impact on the global climate by changing the momentum and energy budgets. Therefore, generation of renewable energy from the wind at this level requires an assessment of the impact on the global climate before such a large amount of wind power will be installed. Such an assessment has to be made with complex Earth system models which are able to simulate the non-linear interactions between the different compartments in the Earth system, i.e. the atmosphere, the biosphere, the hydrosphere, the oceans and the ice.

A first step to address this issue has been made by Wang and Prinn (2010). They have performed simulations with the Community Climate Model Version 3 of the US National Center for Atmospheric Research with a mixed layer ocean (Kiehl et al. 1998) to assess the impact of onshore wind turbines producing 10 % of the global demand in 2010 (4.5 TW or roughly 140 EJ/yr). They find surface warming exceeding 1 °C over onshore wind power installations due to lesser cooling following lower wind speeds within the large wind parks. Significant warming and cooling remote from the installations, and alterations of the global distributions of rainfall and clouds also occur. The climate impacts became negligible when the production fell below 1 TW.

In a second study Wang and Prinn (2011) investigated the effect of offshore wind turbines by increasing the ocean surface drag coefficient. This time they used the Community Atmospheric Model version 3 (CAM3) of the Community Climate System Model (CCSM), developed by the US National Center for Atmospheric Research (NCAR) (Collins et al. 2006). They simulated the impact of installing a sufficient number of wind turbines on coastal waters with depths less than 600 m over the globe that could potentially supply up to 25 % of predicted 2100 world energy needs (45 TW). In contrast to land installation results above (Wang and Prinn (2010), the offshore wind turbine installations are found to cause a surface cooling over the installed offshore regions. This cooling is due principally to the enhanced latent heat flux from the sea surface to lower atmosphere, driven by an increase in turbulent mixing caused by the wind turbines which was not entirely offset by the concurrent reduction of mean wind kinetic energy. Wang and Prinn (2011) found that the perturbation of the large-scale deployment of offshore wind turbines to the global climate is relatively small compared to the case of land-based installations as shown in Wang and Prinn (2010).

A more severe impact of large-scale wind power generation in the order of 10 TW is that such a large extraction of kinetic energy degenerate the efficiency by which the atmosphere converts incoming solar energy into kinetic energy (Miller et al. 2011). Therefore, other forms of renewable energies have to be considered as well for the future energy supply of mankind.

References

Bretton, S.-P., G. Moe: Status, plans and technologies for offshore wind turbines in Europe and North America. Renew. Ener. **34**, 646-654 (2009)

Cañadillas, B., T. Neumann: Comparison Between LES Modelling and Experimental Observations under Offshore Conditions. DEWI Mag. **36**, 48–52 (2010)

Collins, W.D. et al.: The community climate system model version 3 (CCSM3) J. Clim. **19**, 2122–2143 (2006)

Emeis, S.: Measurement Methods in Atmospheric Sciences. In situ and remote. Series: Quantifying the Environment Vol. 1. Borntraeger Stuttgart. XIV+257 pp. (2010b)

Emeis, S.: Surface-Based Remote Sensing of the Atmospheric Boundary Layer. Series: Atmospheric and Oceanographic Sciences Library, Vol. 40. Springer Heidelberg etc., X+174 pp. (2011)

Geng, Q., M. Sugi: Possible Change of Extratropical Cyclone Activity due to Enhanced Greenhouse Gases and Sulfate Aerosols—Study with a High-Resolution AGCM. J. Climate, **16**, 2262–2274 (2003)

Grujicic, M, G. Arakere, B. Pandurangan, V. Sellappan, A. Vallejo, M. Ozen: Multidisciplinary Design Optimization for Glass-Fiber Epoxy-Matrix Composite 5 MW Horizontal-Axis Wind-Turbine Blades. J. Mat. Eng. Perform. **19**, 1116–1127 (2010)

Kiehl, J.T., J.J. Hack, G.B. Bonan, B.A. Boville, D.L. Williams, P.J. Rasch: The National Center for Atmospheric Research Community Climate Model: CCM3. J. Climate, **11**, 1131–1149 (1998)

Miller, L.M., F. Gans, A. Kleidon: Estimating maximum global land surface wind power extractability and associated climatic consequences. Earth Syst. Dynam. **2**, 1–12 (2011)

Nolan, P., P. Lynch, R. McGrath, T. Semmler, S. Wang: Simulating climate change and its effects on the wind energy resource of Ireland. Wind Energy, publ. online 1 Sept 2011, DOI: 10.1002/we.489 (2011)

Steinfeld, G., Tambke, J., Peinke, J., Heinemann, D.: Application of a large-eddy simulation model to the analysis of flow conditions in offshore wind farms. Geophys. Res. Abstr. **12**, EGU2010-8320 (2010)

Thresher, R., M. Robinson, P. Veers: To Capture the Wind. Power and Energy Mag. IEEE, 5, 34–46 (2007)

Trujillo, J.-J., F. Bingöl, G.C. Larsen, J. Mann, M. Kühn: Light detection and ranging measurements of wake dynamics. Part II: two-dimensional scanning. Wind Energy, **14**, 61–75 (2011)

Wang, C., R.G. Prinn: Potential climatic impacts and reliability of very large-scale wind farms. Atmos. Chem. Phys. **10**, 2053–2061 (2010)

Wang, C., R.G. Prinn: Potential climatic impacts and reliability of large-scale offshore wind farms. Environ. Res. Lett. **6**, 025101 (6pp) doi:10.1088/1748-9326/6/2/025101 (2011)

Wichtmann, T., A. Niemunis, T. Triantafyllidis: Validation and calibration of a high-cycle accumulation model based on cyclic triaxial tests on eight sands. Soils Found., **49**, 711–728 (2009)

Yin, J.H.: A consistent poleward shift of the storm tracks in simulations of 21st century climate. Geophys. Res. Lett. **32**, L18701, doi:10.1029/2005GL023684 (2005)

Appenix A
Statistical Tools

This appendix introduces some statistical terms, distributions and techniques which are used throughout the book.

A.1 Time Series Analysis

The advected kinetic energy of an air stream is proportional to the third power of the wind speed, see Eq. (1.1). The climatological mean wind speed is not sufficient to assess the available wind energy at a certain site, because wind turbines can adapt to the actual wind speed within seconds. Additionally, loads and vibrations on structures such as wind turbines depend decisively on the high-frequency parts of the wind spectrum. Therefore, it is important to characterize spatial structures and temporal fluctuations of the wind speed as well. This can be done by computing the wind speed distribution at a site from sufficiently long time series. Time series have to be checked for homogeneity before computing statistical parameters such as those given in Table A.1. Sometimes instruments have been replaced by newer ones at a given measurement site or even the measurement site has been moved to a new position.

For the sake of simplicity and practicability, data distributions are often approximated by mathematical functions that depend on a very low number of parameters. Table A.1 gives an overview of frequently used statistical parameters characterizing the wind.

Frequently, the wind is measured at one point and fluctuations are determined in the time domain. For short time intervals, the "frozen turbulence hypothesis" (also called Taylor's hypothesis) is often used. This hypothesis implies that turbulence elements move with the mean wind and do not alter their shape during such short periods. The frozen turbulence hypothesis allows for a conversion between time and space domain for these short time periods.

S. Emeis, *Wind Energy Meteorology*, Green Energy and Technology,
DOI: 10.1007/978-3-642-30523-8,

Table A.1 Statistical parameters characterizing the wind

Parameter	Description
Mean wind speed	Indicates the overall wind potential at a given site, expected wind speed for a given time interval (first central moment)
Wind speed fluctuation	Deviation of the momentary wind speed from the mean wind speed for a given time interval
Wind speed increment	Wind speed change for a given time span
Variance	Indicates the mean amplitude of temporal or spatial wind fluctuations, expected fluctuation in a given time interval (second central moment)
Standard deviation	Indicates the mean amplitude of temporal or spatial wind fluctuations (square root of the variance)
Turbulence intensity	Standard deviation normalized by the mean wind speed
Gust wind speed	Maximum wind speed in a given time interval
Gust factor	Gust wind speed divided by the mean wind speed in this time interval
Skewness	Indicates the asymmetry of a wind speed distribution around the mean value (third central moment)
Kurtosis (flatness)	Indicates the width of the wind speed distribution around the mean value (fourth central moment)
Excess kurtosis	Kurtosis minus 3
Frequency spectrum	Indicates the frequencies at which the fluctuations occur
Autocorrelation	Indicates the gross spatial scale of the wind speed fluctuations, Fourier transform of the spectrum
Structure function	Indicates the amplitude of wind speed fluctuations, computed from wind speed increments
Turbulent length scale	Indicates the size of the large energy-containing eddies in a turbulent flow
Turbulent time scale	Indicates the time within which wind fluctuations at one point are correlated
Probability density function (pdf)	Indicates the probability with which the occurrence a certain wind speed or wind speed fluctuation can be expected

The time series of the true wind speed $u(t)$ at a given location can be decomposed into a mean wind speed and a fluctuation around this mean (one-point statistics):

$$u(t) = \overline{u(t,T)} + u'(t,T) \tag{A.1}$$

Here, the overbar denotes a temporal average over a time period, T and the prime a deviation from this average. The most frequently used averaging period is 10 min. The mean over the fluctuations is zero by definition:

$$\overline{u'(t,T)} = 0 \tag{A.2}$$

The variance of the time series $u(t)$ is defined as:

$$\sigma_u^2(t,T) = \overline{u'^2(t,T)} \tag{A.3}$$

and the standard deviation is given by the square root of the variance:

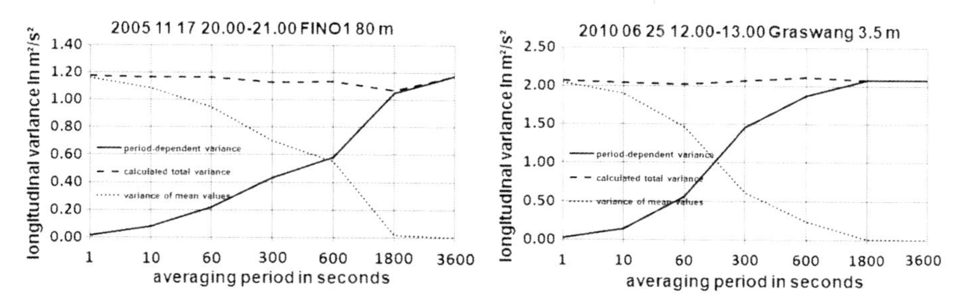

Fig. A.1 Variances measured at 80 m height at FINO1 on November 17, 2005 from 20 to 21 h local time (*left*) and at 3.5 m height at Graswang (Upper Bavaria, Germany) on June 25, 2010 from 12 to 13 h local time (*right*). *Full line* shows variance depending on the averaging period, *dotted line* shows the variance of the mean values and the dashed line gives the total variance following Kaimal et al. (1989)

$$\sigma_u(t,T) = \sqrt{\overline{u'^2(t,T)}} \tag{A.4}$$

Please note that variance and standard deviation depend on the length of the averaging period T as well. Following Kaimal et al. (1989), the variance increases with increasing length of the averaging period. In order to show this, assume that a measurement period can be subdivided into several subperiods. The mean over the whole period is to be denoted by angular brackets and the deviation from this mean by a double prime. A triple prime denotes the deviation of an average over the individual period from the average over the whole period. Then the variance of the deviations from the average over the whole period is the mean of the variances of the individual subperiods plus the variance of the individual mean values from the subperiods:

$$\langle u''^2 \rangle = \left\langle \overline{u'^2} \right\rangle + \langle \overline{u}'''^2 \rangle \tag{A.5}$$

Figure A.1 gives two examples from 10 Hz wind measurements with a sonic anemometer at 80 m at the FINO1 mast in the German Bight (left) and at a rural TERENO site in Graswang (Upper Bavaria, Germany) at 3.5 m height which both prove Kaimal's relation.

The analysis of the increase of the variance with increasing averaging periods can be used to check whether the chosen averaging period is appropriate for the data analysis. The example given on the right-hand side of Fig. A.1 indicates that in this case an averaging period of 1,800 s (30 min) is already sufficient to determine the variance. The example on the left-hand side of Fig. A.1 give hints that even an averaging period of 3,600 s (1 h) might not be sufficient, because the variance is still increasing. If the information of the mean wind speed is available, the period of the strongest increase can be converted into the size of the most energy-containing eddies in the turbulent flow. In the data sample shown in the

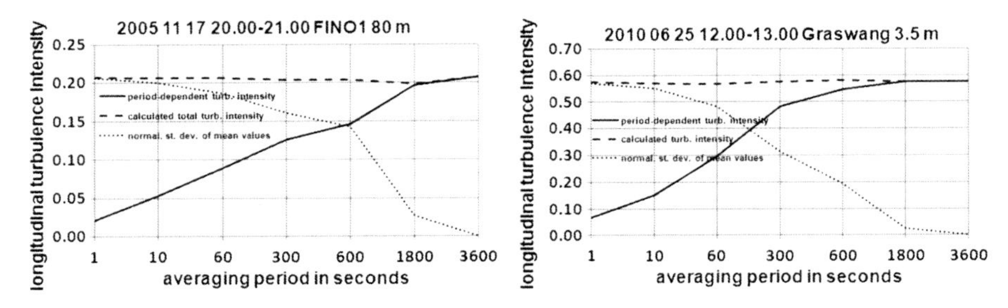

Fig. A.2 Turbulence intensities measured at 80 m height at FINO1 on November 17, 2005 from 20 to 21 h local time (*left*) and at 3.5 m height at Graswang (Upper Bavaria, Germany) on June 25, 2010 from 12 to 13 h local time (*right*). *Full line* shows turbulence intensity depending on the averaging period, *dotted line* shows the standard deviation of the mean values normalized with mean wind speed and the dashed line gives the total turbulence intensity (computed from the total variance in Fig. A.1)

left-hand figure in Fig. A.1 the mean wind speed was 5.2 m/s (mean wind speeds decrease from about 11–4 m/s in the first 15 min of the evaluated time interval and then oscillated around 5.5 m/s for the rest of the period), for the right-hand figure it was 2.5 m/s.

The amplitude of wind fluctuations is usually proportional to the mean wind speed. Therefore, the wind speed variance depends on the mean wind speed in this period. In order to get rid of this dominating wind speed influence, the variance can be normalized with the square of the mean wind speed. Normalization of the standard deviation with the mean wind speed leads to the formation of a frequently used variable: the turbulence intensity. The turbulence intensity is given by:

$$I_u(t,T) = \frac{\sqrt{\overline{u'^2(t,T)}}}{\overline{u(t,T)}} \tag{A.6}$$

Figure A.2 shows the turbulence intensity for the two cases presented in Fig. A.1. Both situations were recorded during unstable thermal stratification. On November 17, 2005, cold air from the North was advected over the still rather warm waters in the German Bight and on June 25, 2010 cool air was present in Upper Bavaria with the sun in a cloudless sky heating the surface considerably. Therefore, turbulence intensities are above average for both marine and land surfaces.

If the wind speed is measured by sonic anemometers rather than by cup anemometers, all three fluctuating components of the wind speed are available for data analysis. Usually, then u' denotes the longitudinal horizontal wind component parallel to the mean wind direction, v' denotes the transverse horizontal wind component perpendicular to the mean wind direction, and w' the vertical wind component. In this case the variances, standard deviations and turbulence intensities can be computed separately for all of these three components. The normalisation of all three components leading to turbulence intensities is done with the mean horizontal wind speed. The variable

$$tke = \frac{\rho}{2}\left(u'^2 + v'^2 + w'^2\right) \tag{A.7}$$

is called turbulent kinetic energy and is a prognostic variable in many numerical flow simulation models.

If wind speed values were distributed totally random, then the probability density function $f(u,T)$ of a stationary time series $u(t)$ for a given averaging period T would follow a normal distribution or Gaussian distribution, which is fully determined by the mean value and the standard deviation:

$$f(u,T) = \frac{1}{\sigma(T)\sqrt{2\pi}} \exp\left(-\frac{1}{2}\left(\frac{u' - \overline{u(T)}}{\sigma(T)}\right)^2\right) \tag{A.8}$$

Higher-order moments can be used to check whether a time series is normally distributed or not. The next two higher moments are skewness:

$$skew_u(t,T) = \frac{\overline{u'^3(t,T)}}{\sigma_u^3(t,T)} \tag{A.9}$$

and flatness or kurtosis:

$$Fl_u(t,T) = \frac{\overline{u'^4(t,T)}}{\sigma_u^4(t,T)} \tag{A.10}$$

The latter is often calculated as excess kurtosis in order to highlight the deviation from normally distributed values which have a flatness of 3:

$$kur_u(t,T) = \frac{\overline{u'^4(t,T)}}{\sigma_u^4(t,T)} - 3 \tag{A.11}$$

For a normal distribution the skewness is zero because this distribution is symmetric. Thus, skewness is a measure for asymmetry of a distribution. Likewise, the excess kurtosis of the normal distribution is zero, because its flatness is 3. Distributions with a negative excess kurtosis have a high central peak and low tails while distributions with a positive excess kurtosis have a lower peak and higher tails.

10 min mean wind speed distributions usually have a positive skewness, i.e., they have a long right tail. This means that large positive deviations from the mean wind speed are more frequent than negative deviations of the same magnitude. This is, because wind speed values are one-sidedly bounded. Negative wind speeds are not possible. Therefore, they cannot be normally distributed and have to be described by other more suitable distributions than the Gaussian distribution, e.g., the two-parametric Weibull distribution (see Sect. A.2 below).

Occurrences of wind speed fluctuations even deviate more from Gaussian statistics. Although the distribution of fluctuations is more or less symmetric around the mean wind speed, analyses have shown that larger wind speed increments (e.g., wind speed changes in 1 or 3 s intervals) are much more frequent than could be expected from Gaussian statistics (see, e.g., Böttcher et al. 2007). Morales et al. (2010) show that only u' values from single 10 min intervals which have been detrended show a distribution close to a normal distribution (excess kurtosis slightly less than zero). u' values from longer time series over many 10 min intervals exhibit an excess kurtosis in the order of 3.3, i.e., large deviations from the mean are much more frequent than it could be expected from a normal distribution. Only normalizing the wind speed deviations u' by the corresponding standard deviation of the respective 10 min interval produces a normal distribution. Motivated by the non-stationarity of atmospheric winds Böttcher et al. (2007) suggest understanding the intermittent distributions for small-scale wind fluctuations as a superposition of different subsets of isotropic turbulence. Therefore, a different statistical approach is necessary. Often, wind fluctuation and gust statistics are described by a Gumbel distribution (Gumbel 1958) which has proven to be especially suitable for extreme value statistics (see Sect. A.3 below).

The fluctuations of wind speed parallel to the mean wind direction (longitudinal component) u', normal to the mean wind direction (transverse component) v' and the vertical component w' are not independent of each other, i.e., they have non-zero correlation products. The most important of these products is:

$$\overline{u'w'} = \frac{1}{T}\int_0^T u'(t)w'(t)dt \tag{A.12}$$

This product is usually negative, because the mean wind speed increases with height and negative (downward) fluctuations of the vertical velocity component bring down positive (higher) longitudinal wind fluctuations from upper layers while positive (upward) fluctuations of the vertical velocity are connected with negative (lower) longitudinal wind fluctuations from lower layers. The square root of the negative value of this correlation product is usually called friction velocity which often serves as a suitable velocity scale in the (mechanically) turbulent atmospheric boundary layer:

$$u_* = \sqrt{-\overline{u'w'}} \tag{A.13}$$

It is a measure how fast horizontal momentum is transported downward by turbulent motions in the atmospheric boundary layer.

Often, one-point statistics are not sufficient to describe the characteristics of atmospheric turbulence. The next step is therefore to look at two-point statistics. A simple example for a two-point statistics in the time domain is the autocorrelation function:

$$R_{u'u'}(\tau) = \frac{1}{\sigma_{u'}^2}\overline{u'(t+\tau)u'(t)} \tag{A.14}$$

where τ is the time lag between the correlated time series. See Fig. A.6 for an example. The autocorrelation function $R(\tau)$ is via a Fourier transformation related to the spectral density of the time series (Morales et al. 2010) and the power spectrum $S(f)$.

$$S(f) = \frac{1}{2\pi} \int_{-\infty}^{\infty} R(\tau)e^{-if\tau}d\tau \tag{A.15}$$

More general two-point statistics can be made by analysing the distributions of wind speed increments δu:

$$\delta u(t,\tau) = u(t+\tau) - u(t) \tag{A.16}$$

The moments of these increments are the structure functions Sf:

$$Sf^n(\tau) = \overline{\delta u(t,\tau)^n} \tag{A.17}$$

Increment probability density functions of wind speed time series are always non-Gaussian (Morales et al. 2010).

A.2 Mean Wind Speed Spectrum and the Weibull Distribution

The wind speed spectrum shows a minimum in the range of about 1 h or ~0.0003 Hz (van der Hoven 1957; Gomes and Vickery 1977; Wieringa 1989). Higher frequencies are usually termed as turbulence. In wind energy this high-frequency turbulence is usually characterized by one variable, the turbulence intensity [see Eq. (A.6) above]. It will be neglected when now looking at frequency distributions for 10 min mean wind speeds, i.e., we will now concentrate on time series of the values $\overline{u(t)}$ which appear as the first term on the right-hand side of the decomposition (A.1). These 10 min mean wind speeds show temporal variations as well. The power spectrum of these low-frequency variations show secondary maxima around 1 day (this is the diurnal variation of the wind), 5–7 days (this is the variation due to the moving weather systems such as cyclones and anticyclones), and around 1 year (the annual variation). The diurnal variation exhibits a phase change with height (Wieringa 1989). The reversal height is roughly at 80 m above ground but values for the reversal height between 40 and 177 m are cited in Wieringa (1989). The phenomenon of the reversal height is closely related to the occurrence of the nocturnal low-level jet in Sect. 3.4. The other long-term variations do not show this phase change.

The frequency distribution of the wind speed for the low-frequency end of the spectrum (i.e., frequencies less than 0.01–0.001 Hz) is usually described by the Weibull distribution. This distribution, which is named after the Swedish engineer, scientist, and mathematician Ernst Hjalmar Waloddi Weibull (1887–1979), is governed by two parameters: a scale factor A (given in m/s, principally

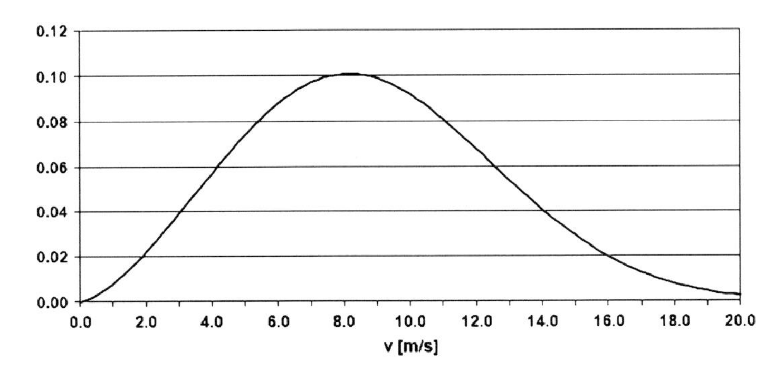

Fig. A.3 Weibull probability density distribution (A.19) for $A = 10$ and $k = 2.5$ as function of wind speed u

proportional to the mean wind speed of the whole time series) and a form factor k (also called shape parameter, dimensionless, describing the shape of the distribution). The probability $F(u)$ of the occurrence of a wind speed smaller or equal to a given speed u is expressed in terms of the Weibull distribution by:

$$F(u) = 1 - \exp\left(-\left(\frac{u}{A}\right)^k\right) \tag{A.18}$$

The respective probability density function $f(u)$ (see Fig. A.3) is found by taking the derivative of $F(u)$ with respect to u:

$$f(u) = \frac{dF(u)}{du} = \frac{k}{A}\left(\frac{u}{A}\right)^{k-1}\exp\left(-\left(\frac{u}{A}\right)^k\right) = k\left(\frac{u^{k-1}}{A^k}\right)\exp\left(-\left(\frac{u}{A}\right)^k\right) \tag{A.19}$$

The mean of the Weibull distribution (the first central moment) and thus the mean wind speed of the whole time series described by the Weibull distribution, $[\overline{u}]$ is given by:

$$[\overline{u}] = A\Gamma(1 + \frac{1}{k}) \tag{A.20}$$

where the square brackets denote the long-term average of the 10 min-mean wind speeds and Γ is the Gamma function. The variance (the second central moment) of this distribution and thus the variance of the 10 min-mean horizontal wind speeds is:

$$\sigma_3^2 = \left[(\overline{u} - [\overline{u}])^2\right] = A^2\left(\Gamma\left(1 + \frac{2}{k}\right) - \Gamma^2\left(1 + \frac{1}{k}\right)\right) \tag{A.21}$$

σ_3^2 is equal to the second term on the right-hand side of (A.5), $\langle \overline{u}'''^2 \rangle$ if the angle brackets defined for that equation denote an average over a day or much longer, and thus become identical with the square brackets. For $k = 1$, the Weibull distribution is equal to an exponential distribution. For $k = 2$, it is equal to the Rayleigh distribution and for about $k = 3.4$ it is very similar to the Gaussian

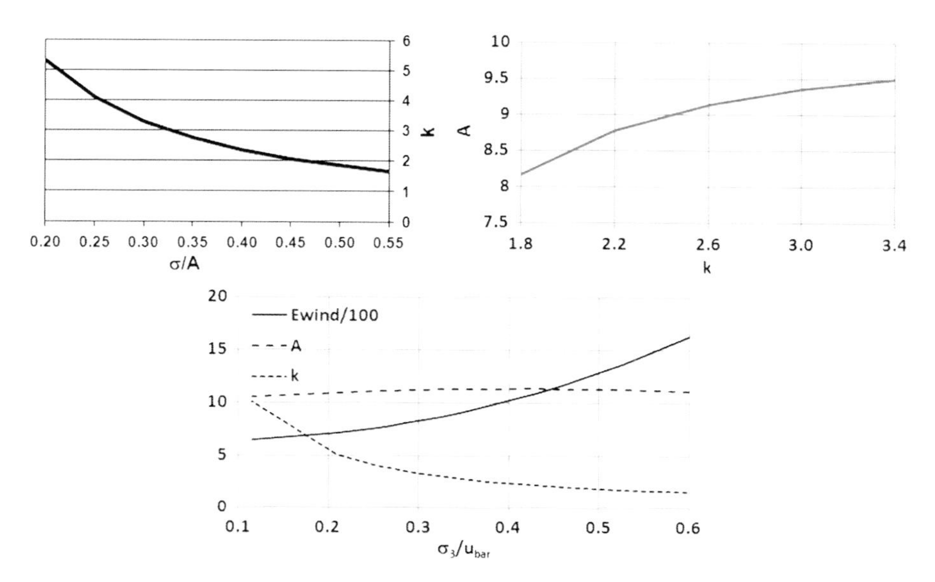

Fig. A.4 *Upper left* Weibull shape parameter k as function of the normalized standard deviation, σ_3/A of the time series. *Upper right* line of equal wind energy. Y-axis: Weibull scale parameter A in m/s, x-axis: Weibull shape parameter k. *Below* energy potential from (A.23) (divided by 100), scale parameter A in m/s and shape parameter k as function of $\sigma_3/[\overline{u}]$ for a mean wind speed $[\overline{u}]$ of 10 m/s

normal distribution. Figure A.3 gives an example for $A = 10$ and $k = 2.5$. The mean value of this sample distribution is 8.87 m/s, the maximum of the distribution is near 8.15 m/s.

Equations (A.20) and (A.21) imply that $[\overline{u}]/A$ as well as σ_3^2/A^2 are functions of k alone. $[\overline{u}]/A$ is only weakly depending on k. It decreases from unity at k = 1 to 0.8856 at k = 2.17 and then slowly increases again. For k = 3, $[\overline{u}]/A$ equals 0.89298. σ_3^2/A^2 is inversely related to k (Wieringa 1989). We find $\sigma_3/A = 1$ for k = 1, $\sigma_3/A = 0.5$ for k = 1.853 and $\sigma_3/A = 0.25$ for k = 4.081 (see also Fig. A.4 upper left).

Higher central moments of the Weibull distribution, M_n are given by (where n is the order of the moment):

$$M_n = A^n \Gamma\left(1 + \frac{n}{k}\right) \tag{A.22}$$

The horizontal flux of kinetic energy of the wind per unit area of the rotor area (usually called wind energy) $E_{wind} = 0.5\ \rho v^3$ is proportional to the third moment of the Weibull distribution and can be easily calculated once A and k are known:

$$E_{wind} = 0.5\rho A^3 \Gamma\left(1 + \frac{3}{k}\right) \tag{A.23}$$

As the relation between the mean wind speed (A.20) and the wind energy (A.23) is non-linear, different combinations of A and k can lead to the same mean wind energy (see Fig. A.4 upper right for an example). Likewise, for a given mean wind speed, wind energy from (A.23) increases with an increasing variation of the wind speed $\sigma_3/[\overline{u}]$ (see Fig. A.4 bottom). Thus, for a correct estimation of the wind energy, the parameters A and k have to be known, not just the mean wind speed.

For a practical determination of the two Weibull parameters A and k from a time series of wind speed values, we take the double logarithm of the relation (A.18) following Justus et al. (1976):

$$\begin{aligned} y &= \ln\left\{\ln\left\{1 - \left(1 - \exp\left(-\left(\frac{u}{A}\right)^k\right)\right)\right\}\right\} = \ln\left\{\ln\left\{\exp\left(-\left(\frac{u}{A}\right)^k\right)\right\}\right\} \\ &= k\ln A - k\ln u = a + b\ln u \end{aligned} \tag{A.24}$$

From (A.24) A and k can be determined by fitting a straight line into a plot of y against ln u. We get the scale factor A from the intersection a of the fitted line with the y-axis:

$$A = \exp\left(\frac{a}{k}\right) \tag{A.25}$$

and the form factor k from the negative slope b of this line:

$$k = -b \tag{A.26}$$

Inversion of (A.20) and an exponential fit to (A.21) gives alternatively (Justus et al. 1978) a useful relation between A and k and the mean wind speed $[\overline{u}]$ and the standard deviation σ_3:

$$A = \frac{[\overline{u}]}{\Gamma(1 + \frac{1}{k})} \tag{A.27}$$

and

$$k = \left(\frac{\sigma_3}{[\overline{u}]}\right)^{-1.086} \tag{A.28}$$

Relation (A.28) is plotted in the lower frame of Fig. A.4 for a constant value of $[\overline{u}]$. Sensitivity calculations show that the wind energy estimate from (A.23) is much more sensitive to the correct value of A than to the value of k. An uncertainty in A of 10 % leads to a deviation of 30 % in the estimated wind energy. An uncertainty in k of 10 % on the other hand only leads to a deviation of 9 % in the estimated wind energy. An overestimation of k yields an underestimation of the wind energy and vice versa.

Please note that $\sigma_3/[\overline{u}]$ in (A.28) is different from the turbulence intensity I_u defined in (A.6). Usually $\sigma_3/[\overline{u}]$ is considerably larger than I_u, because it represents the much larger diurnal, synoptic and seasonal fluctuations of the 10 min-mean wind speeds, while I_u describes the smaller short-term fluctuations during a 10 min interval after any longer trends and variations have been subtracted from the data in this interval. Equation (A.28) can be inverted in order to estimate the order of magnitude of $\sigma_3/[\overline{u}]$. This ratio is of the order of 1/k, i.e., 0.4–0.5 while the turbulence intensity over land is in the order of 0.2 and the offshore turbulence intensity is usually below 0.1.

A.3 Extreme Mean Wind Speeds and the Gumbel Distribution

Extreme mean wind speeds are important for load estimations for wind turbines. Usually, they have to be specified for a certain return period which is related to the time period for which the turbine is expected to operate. The probability of occurrence of extreme values can be described by a Gumbel distribution (Gumbel 1958). This distribution is a special case of a generalized extreme value distribution or Fisher-Tippett distribution as is the Weibull distribution (Cook 1982; Palutikof et al. 1999). It is named after the German mathematician Emil Julius Gumbel (1891–1966).

The probability density function for the occurrence of a largest value x reads:

$$f(x) = e^{-x}e^{-e^{-x}} \tag{A.29}$$

Due to its form this distribution is often call double exponential distribution. The related cumulative frequency distribution reads:

$$F(x) = e^{-e^{-x}} \tag{A.30}$$

The inverse of (A.30) is the following percent point function:

$$G(p) = -\ln(-\ln(p)) \tag{A.31}$$

The 98th percentile ($p = 0.98$) of this percent point function has the value 3.9, the 99th percentile the value 4.6, and the 99.9th percentile the value 6.9.

The practical calculation from a given time series may be done as follows: In a first step, independent maxima of a wind speed time series (e.g., annual extreme values) are identified. Then, these maxima are sorted in ascending order forming a new series of maxima with N elements. The cumulated probability p that a value of this new series is smaller than the mth value of this series is $p(m) = m/(N + 1)$. Finally, the sorted values are plotted against the double negative logarithm of their cumulative probability, i.e., they are plotted against $-\ln(-\ln(p))$. Data which follow a Gumbel distribution organize along a straight line in such a graph. Once the graph is plotted, estimations of extreme values for a given return period are easy. For example, from a statistics of annual extreme values, u_{max} the extreme value which is expected to appear once in 50 years is found where the extrapolated straight line

$$u_{\max} = a(-\ln(-\ln(p))) + b \tag{A.32}$$

crosses the value 3.9 ($p = 1-1/50 = 0.98$ and $-\ln(-\ln(0.98)) = 3.9$).

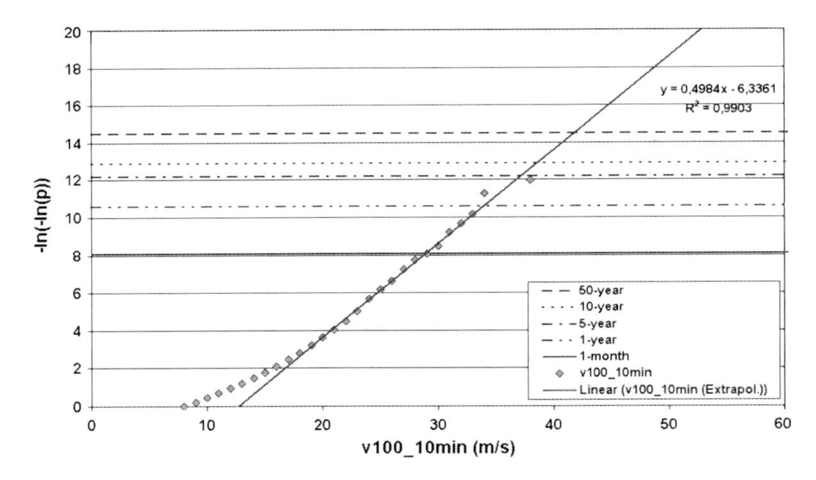

Fig. A.5 Gumbel plot of a time series from 10 min mean wind speeds observed at 100 m at the FINO 1 mast in the German Bight during the 4 years from September 2003 to August 2007. Wind data have been lumped into 1 m/s bins

If a time series is much shorter than the interesting return period, then the series of annual extreme values will be too short for a meaningful analysis. For example, it does not make sense to pick out four annual extreme values from a four-year time series and to extrapolate a straight line through this data. But another possibility exists in this case, which has been demonstrated in Emeis and Türk (2009). Here, the 50 year extreme mean wind speed had been estimated from 4 years of 10 min mean wind data (about 200,000 data points). This procedure has also been used in Carter (1993) and Panchang et al. (1999) and is based on the assumption that the wind speed time series follows a Fisher-Tippett Type 1 distribution.

The probability of a 50 year extreme from such a time series with 10 min intervals (52,560 data points a year) is given by $p = 1-1/(50 \times 52{,}560)$, giving $-\ln(-\ln(p)) = 14.78$. For hourly values the threshold value would be 12.99.

Figure A.5 shows the Gumbel plot of a wind speed time series based on 10 min mean values. It features a nearly perfect straight line for the extreme wind speeds above 18 m/s (the large majority of values are below 18 m/s and these follow a Weibull distribution which does not give a straight line in a Gumbel plot). The equation for this straight line according to (A.32) is $u_{max} = 2.01\ (-\ln(-\ln(p))) + 12.71$ (the inverse of this equation is given in the upper right of Fig. A.5). This can be used to extrapolate to the 50 year extreme value of the 10 min-average wind speed which turns out be $2.01 \times 14.78 + 12.71 = 42.42$ m/s in this example.

A.4 Extreme Gusts

Extreme wind gusts lead to short-term loads on wind turbines. The gust wind speed, u_{gust} can be coupled to the mean wind speed via a gust factor G. This factor usually depends on the averaging time for the gust t, the related averaging time for the mean wind speed T ($t \ll T$), the height above ground z, the surface roughness z_0 and the mean wind speed $\overline{u}$ (Wieringa 1973; Schroers et al. 1990):

$$G(t, T, z, z_0, \overline{u}) = \frac{u_{gust}(t, z)}{\overline{u(T, z, z_0)}} \tag{A.33}$$

Trends have to be removed before calculating G (Wieringa 1973). Vertical profiles of G are discussed in Chap. 3. Frequency distributions of G can be described by a Weibull distribution (Jensen and Kristensen 1989). Assuming a normal distribution of the momentary wind speeds in an averaging interval (which probably is a good assumption for higher wind speeds), i.e., stipulating:

$$u_{gust}(t, z) = \overline{u(T, z, z_0)} + k\sigma_u(T, t, z, z_0, \overline{u}) \tag{A.34}$$

allows for a description of the gust factor G from the standard deviation and the mean wind speed (Mitsuta and Tsukamoto 1989):

$$G(t, T, z, z_0, \overline{u}) = 1 + \frac{k\sigma_u(T, t, z, z_0, \overline{u})}{\overline{u(T, z, z_0)}} = 1 + kI_{\overline{u}}(t, T) \tag{A.35}$$

where k is a so-called peak factor. Equation (A.35) shows, that the gust factor is closely related to the turbulence intensity [see (A.6)]. Wieringa (1973) gives for k:

$$k(D/t) = 1.42 + 0.3013\ln(D/t - 4) \tag{A.36}$$

while Mitsuta and Tsukamoto (1989) cite a more simple relation:

$$k(D/t) = (2\ln(D/t))^{0.5} \tag{A.37}$$

where D is the length of the observation period. Typical values for G are in the order of 1.3–1.4. Wieringa (1973) showed that gust factors for hourly mean wind speeds are about 10 % higher than for 10 min mean wind speeds. Over land G usually decreases with increasing wind speed due to the similar behaviour of σ_u and I_u (Davis and Newstein 1968).

The Gumbel method presented in Sect. A.3 can be used to estimate a 50 year extreme 1 s gust as well. An evaluation from the FINO1 dataset from September 2003–August 2007 gives 52.1 m/s (Türk 2008).

A.5 Gust Duration and Wind Acceleration in Gusts

Gusts are characterized by a rapid increase in wind speed and a subsequent decrease. For load estimations, a so-called "Mexican hat" shape of the gust (see, e.g., Fig. 7.29 below) is assumed (e.g., in the standard IEC 61400-1) for wind turbine load calculations, which starts with a wind speed decrease before the rapid increase and a similar overshooting for the wind speed decrease directly afterwards [see (A.39)]. The maximum expected gust amplitude over the rotor-swept area is assumed to be:

$$u_{gust} = \min\left\{1.35(u_{e1} - u),\ 3.3\left(\frac{\sigma_u}{1 + 0.1\left(\frac{D}{\Lambda_1}\right)}\right)\right\} \tag{A.38}$$

where u_{e1} is the extreme 3 s gust with a recurrence period of 1 year, D is the rotor diameter in m and Λ_1 is a turbulent length scale parameter, which in IEC 61400-1 is put to 42 m for larger wind turbines with hub heights above 60 m The time variation of wind speed in such a "Mexican hat" gust event is assumed to be (see Fig. 5.29 for an example):

$$u(t) = \begin{cases} u - 0,37u_{gust}\sin(3\pi t/T)(1 - \cos(2\pi t/T)) & for\ 0 \le t \le T \\ u & otherwise \end{cases} \tag{A.39}$$

where T is assumed to be 10.5 s. This "Mexican hat" model implies an increase of wind speed from the lowest value to the highest value in about 4 s. Investigations at the FINO1 platform in the German Bight (Türk 2008) have shown that gusts with a time period of 8 s are even more frequent than those with 10.5 s (see Sect. 5.4 for more details).

A.6 Size of Turbulence Elements

The size of turbulent elements depends on the distance to the surface underneath, as this distance is a limiting factor for the growth of these elements. One method to estimate the size of turbulent elements in a turbulent flow is to analyse the autocorrelation function (A.12). The integral over the autocorrelation function to the first zero crossing of the autocorrelation function indicates the longitudinal time scale T_u of a turbulence element at a given position.

$$T_u = \int_0^\infty R_{u'u'}(\tau)d\tau \tag{A.40}$$

Because the autocorrelation function is usually an exponential function (Foken 2008), this time scale can be approximated as the time lag, τ at which the

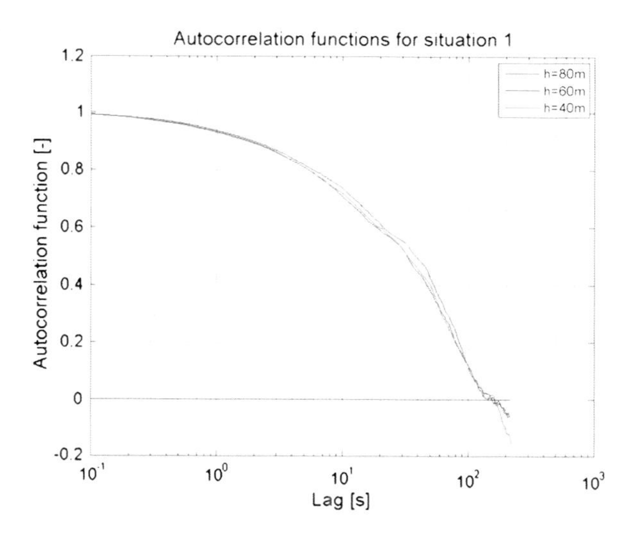

Fig. A.6 Typical example of an autocorrelation function from FINO1 data (Fig. 3.3 from Türk 2008)

autocorrelation function has decreased to $1/e \approx 0.37$. Multiplying the time scale with the mean wind speed u gives the spatial dimension Λ_u of this element in flow direction und the assumption of the validity of Taylor's frozen turbulence hypothesis. This spatial dimension is called integral turbulence length scale.

$$\Lambda_u = \overline{u}T_u = \overline{u}\int_0^\infty R_{u'u'}(\tau)d\tau \tag{A.41}$$

Figure A.6 shows an example where the time scale is of the order of 50 s. As the mean wind speed increases with height, this means that the integral turbulence length scale according to (A.40) increases with height as well.

Appendix B
Remote Sensing of Boundary Layer Structure and Height

The mixed layer height (MLH) and the boundary layer height appear as height scales in several approaches for the description of vertical wind and turbulence profiles in Chap. 3. MLH is the height up to which atmospheric properties or substances originating from the Earth's surface or formed within the surface layer are dispersed almost uniformly over the entire depth of the mixed layer by turbulent vertical mixing processes. Therefore, the existence and the height of a mixed layer can either be analyzed from detecting the presence of the mixing process, i.e., turbulence, or from the verification that a given conservative atmospheric variable is distributed evenly over a certain height range. The level of turbulence can for instance be derived from fluctuations of the wind components or from temperature fluctuations. Suitable conservative atmospheric variables for the identification of the mixed layer and its height are, e.g., potential temperature, specific humidity or aerosol particle concentrations (Fig. B.1).

Figure B.1 shows two snapshots from a diurnally varying boundary layer under clear-sky conditions as depicted in Fig. 3.2. The left frame in Fig. B.1 is valid around noon, the right frame around midnight. For a convective boundary layer at noon, MLH and boundary-layer height are more or less identical, the vertical mixing is thermally driven and reaches right to the top of the boundary layer. At night, when mechanically produced turbulence is present only, MLH is usually identical with the much lower height of the stable surface layer. The nocturnal boundary-layer height is usually identical to the top height of the residual layer, which is a remnant from the daytime convective layer. Because MLH is not a primary atmospheric variable and cannot be determined from in situ measurements at the surface, this Appendix B has been added here in order to illustrate the necessary measurement efforts to determine this parameter. Figure B.1 shows that distinct features in the vertical profiles of atmospheric turbulence, temperature, specific humidity and the aerosol content are appropriate to determine the mixed layer height and the boundary layer height.

In situ measurements of the abovementioned conservative variables can be done over the necessary vertical height range of up to 1 or 2 km only by launching

S. Emeis, *Wind Energy Meteorology*, Green Energy and Technology,
DOI: 10.1007/978-3-642-30523-8, © Springer-Verlag Berlin Heidelberg 2013

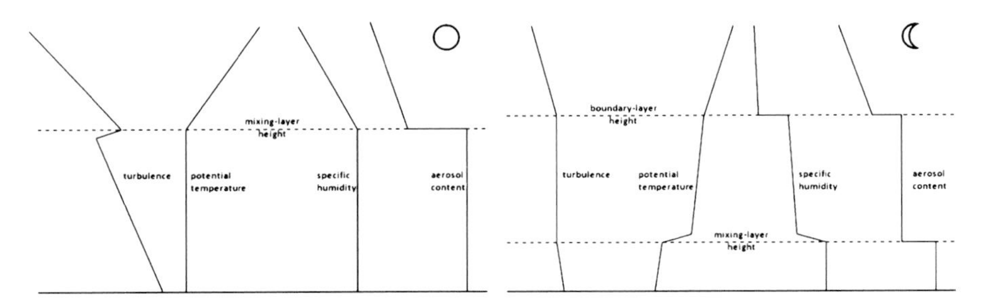

Fig. B.1 Principal sketch of vertical profiles of some important variables within the well-mixed daytime atmospheric boundary layer (*left*) and the more stable nocturnal surface layer and the residual layer (lower and middle layer in the *right-hand* frame) and above in the free troposphere

radiosondes. Evaluation of radiosonde data gives quite reliable data in most cases. The great disadvantage of radiosondes is the missing temporal continuity. Therefore, remote sensing methods are preferable although (with the exception of RASS which directly detect temperature profiles) they only give an indirect detection of the mixing height. A first rather complete overview of methods to determine the MLH from in situ measurements and surface-based remote sensing has been given by Seibert et al. (2000). Since then considerable development has taken place, especially with regard to the usage of surface-based remote sensing methods [see the review paper by Emeis et al. (2008) and the monograph by Emeis (2011)]. This Appendix will mainly follow these sources.

Newly developed optical methods for MLH detection illustrate this recent progress. Seibert et al. (2000) still classified LIDAR methods as expensive, not eye-save, with a high lowest range gate, limited range resolution, and sometimes subject to ambiguous interpretation. This has changed drastically in the last 10 years when better and smaller LIDARs have been built and ceilometers have been discovered to be a nearly ideal boundary layer sounding instrument. Progress has been made in the field of acoustic sounding as well. Similarly, algorithms for the determination of MLH from vertical profiles of the acoustic backscatter intensity as described in Beyrich (1997) and Seibert et al. (2000) have been enhanced by using further variables available from SODAR measurements such as the wind speed and the variance of the vertical velocity component (Asimakopoulos et al. 2004; Emeis and Türk 2004). Such enhancements had been named as possible methods in Beyrich (1995) and Seibert et al. (2000) but obviously no example was available at that time.

A variety of different algorithms have been developed by which the MLH is derived from ground-based remote sensing data (see Table B.1 for a short overview). We will mainly concentrate on acoustic and optical remote sensing because electro-magnetic remote sensing with wind profilers has too high lowest range gates for a good coverage of shallow MLH. The disadvantage of a too high lowest range gate can sometimes partly be circumvented by slantwise profiling or conical scanning if the assumption of horizontal homogeneity can be made.

Table B.1 Overview on methods using ground-based remote sensing for the derivation of the mixed layer height mentioned in this Appendix (see rightmost column for section number in this Appendix)

Method	Short description	Section
Acoustic ARE	Analysis of acoustic backscatter intensity	B.1.1
HWS	Analysis of wind speed profiles	B.1.2
VWV	Analysis of vertical wind variance profiles	B.1.3
EARE	Analysis of backscatter and vertical wind variance profiles	B.1.4
Optical threshold	Detection of a given backscatter intensity threshold	B.2.1
Gradient	Analysis of backscatter intensity profiles	B.2.2
Idealised backscatter	Analysis of backscatter intensity profiles	B.2.3
Wavelet	Analysis of backscatter intensity profiles	B.2.4
Variance	Analysis of backscatter intensity profiles	B.2.5
Acoust./electro-magn.	RASS	B.3
	SODAR-RASS and windprofiler-RASS	B.3.1
/in situ	SODAR-RASS plus surface heat flux data	B.3.2
Acoust./electro-magn.	SODAR plus windprofiler	B.4.1
Acoustic/optical	SODAR plus ceilometer	B.4.2

B.1 Acoustic Detection Methods

Acoustic methods for the determination of MLH either analyze the acoustic backscatter intensity, or, if Doppler shifts in the backscattered pulses can be analyzed, features of vertical profiles of the wind components and its variances as well. The acoustic backscatter intensity is proportional to small-scale fluctuations in atmospheric temperature (usually generated by turbulence) or by stronger vertical temperature gradients. The latter feature may be an indication for the presence of temperature inversions, which can often be found at the top of the mixed layer (Fig. B.1).

Beyrich (1997) listed possible analyses which can mainly be made from acoustic backscatter intensities measured by a SODAR. Later, Asimakopoulos et al. (2004) summarized three different methods to derive MLH from SODAR data: (1) the horizontal wind speed method (HWS), (2) the acoustic received echo method (ARE), and (3) the vertical wind variance method (VWV). We will mainly follow this classification here and finally add a fourth method, the enhanced ARE method (EARE), in Sect. B.1.4.

B.1.1 Acoustic Received Echo Method

The acoustic received echo method (ARE) is the oldest and most basic method of determining MLH from acoustic remote sensing. Most of the methods listed in

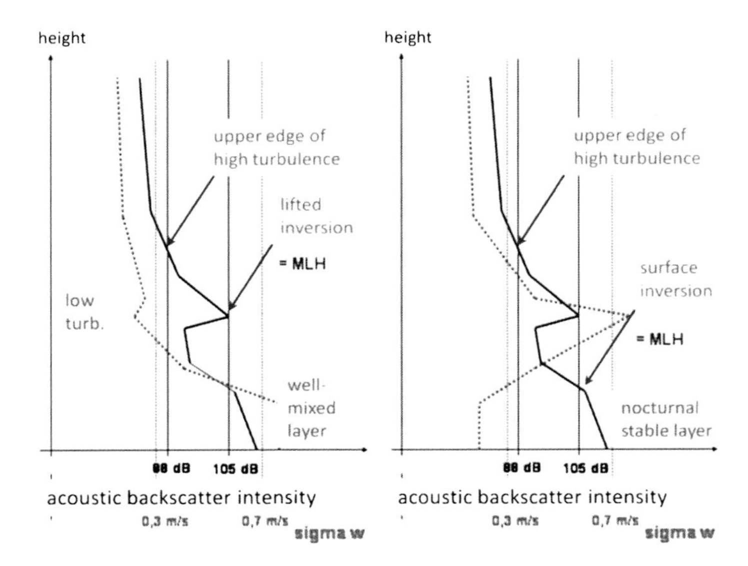

Fig. B.2 Principal sketch of the EARE method for determining the mixed layer height from vertical profiles of the acoustic backscatter intensity (*bold line*) and the variance of the vertical velocity component (sigma w in m/s, *dotted line*). *Left* lifted inversion, *right* stable nocturnal layer with very low mixed layer height

Beyrich (1997) belong to this method. The method does not require a Doppler shift analysis of the backscattered signals but is based on the analysis of facsimile plots, i.e., time-height cross-sections of the backscatter intensity. The method makes use of the assumption that turbulence is larger in the mixed layer than in the atmosphere above, and that this turbulence is depicted in enhanced intensity of the acoustic backscatter. MLH is analyzed either from the maximum negative slope or from the changing curvature of the vertical profile of the acoustic backscatter intensity or it is analyzed from the height where the backscatter intensity decreases below a certain pre-specified threshold value.

B.1.2 Horizontal Wind Speed Method

The horizontal wind speed method (HWS) requires a Doppler shift analysis of the backscattered acoustic signals. The algorithm is based on the analysis of the shape of hourly-averaged vertical wind speed profiles using the assumption that wind speed and wind direction are almost constant within the mixed layer but approach gradually towards the geostrophic values above the mixed layer. Beyrich (1997) listed this method in his Table 2 but did not discuss it further. The applicability of the method is probably limited to well-developed convective boundary layers (CBL) due to the underlying assumptions. Such CBL are often higher than the maximum range of a SODAR. Even if the CBL height is within the range of the SODAR the algorithm for the analysis of the Doppler shift often fails above the inversion topping the CBL due to too low signal-to-noise ratios. Today, small Doppler wind lidars are available to derive wind speed and direction profiles through the whole depth of the boundary layer. This facilitates the application of the HWS method.

B.1.3 Vertical Wind Variance Method

The vertical wind variance method (VWV) is also working only for CBLs. It is based on the vertical profile of the variance of the vertical velocity component σ_w. In a CBL σ_w reaches a maximum in a height $a \cdot z_i$. Typical values for a are between 0.35 and 0.4. Thus, in principle, this is an extrapolation method. It has been applied to SODAR measurements because it permits a detection of MLH up to heights which are 2.5 times above the limited maximum range (usually between 500 and 1,000 m) of the SODAR. Beyrich (1997) classified this method as not reliable. A related method, which is based on power spectra of the vertical velocity component, is integrated in the commercial evaluation software of certain SODARs (Contini et al. 2009). The application of the VWV method is now also been facilitated by the easy availability of small Doppler wind lidars.

B.1.4 Enhanced Acoustic Received Echo Method

The enhanced acoustic received echo method (EARE) algorithm is an extension of the ARE method and has been proposed by Emeis and Türk (2004) and Emeis et al. (2007b). It includes the variance of the vertical velocity component into the MLH algorithm which is available from Doppler-SODAR measurements. Additionally, it does not only determine the MLH but also the heights of additional lifted inversions. Especially in orographically complex terrain, the vertical structure of the ABL can be very complicated. Emeis et al. (2007a) have shown that several persistent inversions one above the other which form in deep Alpine valleys can be detected from SODAR measurements.

EARE determines three different types of heights based on acoustic backscatter intensity and the variance of the vertical velocity component (see Fig. B.2). Because the horizontal wind information above the inversion is not regularly available from SODAR measurements, horizontal wind data have not been included into this scheme. In the following a letter "H" and an attached number will denote certain derived heights which are related to inversions and the MLH; while the variable z is used to denote the normal vertical coordinate. The EARE algorithm detects:

- the height ($H1$) of a turbulent layer characterised by high acoustic backscatter intensities $R(z)$ due to thermal fluctuations (therefore having a high variance of the vertical velocity component σ_w),
- several lifted inversions ($H2\_n$) characterized by secondary maxima of acoustic backscatter due to a sharp increase of temperature with height and simultaneously low σ_w, and
- the height of a surface-based stable layer ($H3$) characterised by high backscatter intensities due to a large mean vertical temperature gradient starting directly at the ground and having a low variance of the vertical velocity component.

The height $H1$ corresponds to a sharp decrease $\partial R/\partial z < DR_1$ of the acoustic backscatter intensity $R(z)$ below a threshold value R_c with height z usually indicating the top of a turbulent layer. $R_c = 88$ dB and $DR_1 = -0.16$ dB/m have proven to be meaningful values in the abovementioned studies. R_c is somewhat arbitrary because the received acoustic backscatter intensities from a SODAR cannot be absolutely calibrated. An absolute calibration would require the knowledge of temperature and humidity distributions along the sound paths for a precise calculation of the sound attenuation in the air. DR_1 is, at least for smaller vertical distances, independent from the absolute value of R_c. An application-dependent fine-tuning of R_c and DR_1 may be necessary.

Elevated inversions are diagnosed from secondary maxima of the backscatter intensity that are not related to high turbulence intensities. For elevated inversions increase in backscatter intensity below a certain height $z = H2$ and a decrease above is stipulated while the turbulence intensity is low. The determination of the height of the stable surface layer $H3$ is started if the backscatter intensity in the lowest range gates is above 105 dB while σ_w is smaller than 0.3 ms^{-1}. The top of the stable layer $H3$ is at the height where either the backscatter intensity sinks below 105 dB or σ_w increases above 0.3 ms^{-1}. The threshold values for σ_w have been determined by optimizing the automatic application of the detection algorithm. In doing so it turned out that no lifted inversions occurred with a variance σ_w higher than 0.7 ms^{-1} and that the variance σ_w in nocturnal stable surface layers was below 0.3 ms^{-1}. The first σ_w threshold made it possible to distinguish between inversions and elevated layers of enhanced turbulence. The latter σ_w threshold made it possible to differentiate between nocturnal stable surface layers and daytime super-adiabatic surface layers although both types of surface layers yield more or less the same level of backscatter intensity. Finally MLH from the acoustic remote sensing is determined as the minimum of $H1$, $H2\_1$, and $H3$.

B.2 Optical Detection Methods

Usually the particle content of the mixed layer is higher than in the free troposphere above (Fig. B.1), because the emission sources for aerosol particles are in most cases at the ground. Particle formation from precursors mainly takes place near the surface as well. Making the assumption that the vertical particle distribution adapts rapidly to the changing thermal structure of the boundary layer, MLH can be determined from the analysis of the vertical aerosol distribution. This also includes the assumption that the vertical aerosol distribution is not dominated by horizontally advected aerosol plumes or layers. The heights of near surface aerosol layers (*H4_n*) can be analysed from optical vertical backscatter profiles obtained by optical remote sensing. Several methods have been developed, the most prominent of these being: (1) the threshold method, (2) the gradient or derivative method, (3) the idealised gradient method, (4) the wavelet method, and (5) the variance method. In addition, the abovementioned horizontal wind speed method (Sect. B.1.2) and vertical wind variance method (Sect. B.1.3) are available to derive the vertical structure of the boundary layer from Doppler wind lidar data.

The application of optical remote sensing for MLH determination has focussed on the use of ceilometers in recent years but small wind lidars usually provide this information as well. In contrast to wind lidars, ceilometers do not determine the Doppler shift of the backscattered signal. For the detection of MLH below 150–200 m a ceilometer with one optical axis for the emitted and the received beam should be used. Due to the thin light beams the overlap of the emitted and received beam from a ceilometer with two parallel optical axes can be insufficient in this height range.

B.2.1 Threshold Method

Melfi et al. (1985) and Boers et al. (1988) used simple signal threshold values, though this method suffers from the need to define them appropriately (Sicard et al. 2006). *H4* is defined here as the height within the vertical profile of the optical backscatter intensity where the backscatter intensity first exceeds a given threshold when coming downward from the free unpolluted troposphere. The determination of several heights *H4_n* would require the definition of several thresholds which probably cannot be done a priory to the analysis. Therefore this will always lead to a subjective analysis of MLH.

B.2.2 Gradient or Derivative Methods

Hayden et al. (1997) and Flamant et al. (1997) proposed to use the largest negative peak of the first derivative of the optical attenuated backscatter intensity ($B(z)$) for the detection of *H4* from LIDAR data (height of gradient minimum $H4_{GM}$):

$$H4_{GM} = \min(\partial B(z)/\partial z) \qquad \text{(B.1)}$$

Likewise Wulfmeyer (1999) used the first minimum of the slope to detect the top of a convective boundary layer from DIAL data. Münkel and Räsänen (2004),

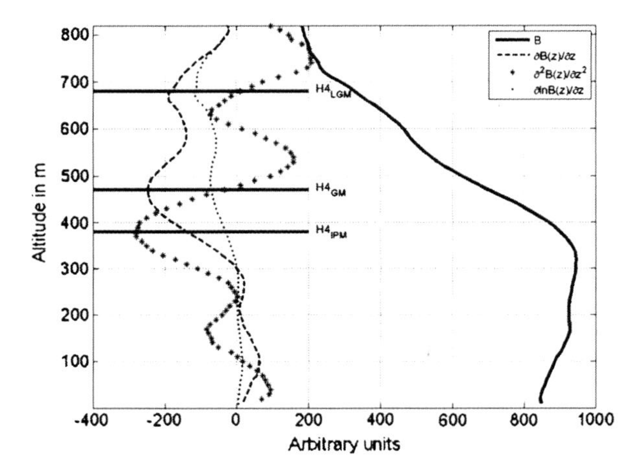

Fig. B.3 Comparison of three different methods [see Eqs. (B.1)–(B.3)] determining the mixed layer height from optical backscatter intensity [from Emeis et al. (2008)]

Münkel (2007), and Schäfer et al. (2004, 2005) applied the gradient method to ceilometer data. Menut et al. (1999) took the minimum of the second derivative of $B(z)$ as the indication for MLH:

$$H4_{IPM} = \min\left(\partial^2 B(z)/\partial z^2\right) \tag{B.2}$$

This method is called inflection point method (IPM). It usually gives slightly lower values for $H4$ than the gradient method (B.2). A further approach was suggested by Senff et al. (1996). They looked for the largest negative gradient in the logarithm of the backscatter intensity (height of logarithmic gradient minimum $H4_{LGM}$):

$$H4_{LGM} = \min(\partial \ln B(z)/\partial z) \tag{B.3}$$

This approach usually gives the largest value for $H4$. According to Sicard et al. (2006) $H4_{IPM}$ from (B.2) is closest to the MLH derived from radiosonde ascents via the Richardson method. The other two algorithms (B.1) and (B.3) give slightly higher values. The vertical profiles shown in Fig. B.3 (taken from Emeis et al. 2008) give a comparison of the determination of mixed layer heights from Eqs (B.1) to (B.3).

In Emeis et al. (2007a) the gradient method (B.1) has been further refined and extended to enable the calculation of up to $n = 5$ lifted inversions. Prior to the determination of gradient minima the overlap and range corrected attenuated backscatter profiles have to be averaged over time and height to suppress noise generated artefacts. Therefore the $H4$ values are determined in a two-step procedure. Between 140 and 500 m height sliding averaging is done over 15 min and a height interval Δh of 80 m. In the layer between 500 and 2,000 m Δh for vertical averaging is extended to 160 m. Two additional parameters have been introduced to further reduce the number of false hits. The minimum accepted attenuated backscatter intensity B_{min} right below a lifted inversion is set to

200×10^{-9} m^{-1}sr^{-1} in the lower layer and 250×10^{-9} m^{-1}sr^{-1} in the upper layer. Additionally the vertical gradient value $\partial B/\partial z_{max}$ of a lifted inversion must be more negative than -0.30×10^{-9} m^{-2}sr^{-1} in the lower layer and more negative than -0.60×10^{-9} m^{-2}sr^{-1} in the upper layer.

B.2.3 Idealised Backscatter Method

A parallel development by Eresmaa et al. (2006) using an idealised backscatter profile, originally described by Steyn et al. (1999), is also an extension of the gradient method. MLH is not determined from the observed backscatter profile, but from an idealised backscatter profile fitted to the observed profile. The robustness of this technique is founded on utilising the whole backscatter profile rather than just the portion surrounding the top of the mixed layer. In this method an idealized backscattering profile $B_i(z)$ is fitted to measured profile by the formula:

$$B_i(z) = \frac{B_m + B_u}{2} - \frac{B_m - B_u}{2} erf\left(\frac{z - h}{\Delta h}\right) \tag{B.4}$$

where B_m is the mean mixed layer backscatter, B_u is the mean backscatter in air above the mixed layer and Δh is related to the thickness of the entrainment layer capping the ABL in convective conditions.

B.2.4 Wavelet Method

A wavelet method has been developed for the automatic determination of mixed layer height from backscatter profiles of an LD-40 ceilometer by de Haij et al. (2006). Before that wavelet transforms have been applied in recent studies for MLH determination from LIDAR observations (e.g., Cohn and Angevine 2000; Davis et al. 2000; Brooks 2003; Wulfmeyer and Janjić 2005). The most important advantage of wavelet methods is the decomposition of the signal in both altitude as well as vertical spatial scale of the structures in the backscatter signal.

The wavelet algorithm in de Haij et al. (2006) is applied to the 10 min averaged range and overlap corrected backscatter profile $B(z)$ within a vertical domain of 90–3,000 m. For each averaged profile the top of two significant aerosol layers are detected in order to detect MLH as well as the top of a secondary aerosol layer, like e.g., an advected aerosol layer or the residual layer. This wavelet MLH method uses the scale averaged power spectrum profile $W_B(z)$ of the wavelet transform with 24 dilations between 15 and 360 m and step size 15 m. The top of the first layer, *H4_1*, is detected at the first range gate at which the scale averaged power spectrum $W_B(z)$ shows a local maximum, exceeding a threshold value of 0.1. This threshold value is empirically chosen, based on the analysis of several cases with both well pronounced and less clearly pronounced mixed layer tops. *H4_2* is optionally determined in the height range between *H4_1* and the upper boundary of detection. A valid *H4_2* is detected at the level with the strongest local maximum of $W_B(z)$ provided that this maximum is larger than the $W_B(z)$ of *H4_1*. MLH is set equal to *H4_1*.

However, problems with this method arise e.g., in case of multiple (well defined) aerosol layers, which renders the selection of the correct mixed layer top ambiguous. Furthermore, in spring and summer the detection of the MLH for deep (convective) boundary layers often fails. This is mostly due to the high variability of the aerosol backscatter signal with height which limits the range for MLH estimation in those conditions (de Haij et al. 2006).

B.2.5 Variance Method

At the top of the convective boundary layer (CBL) we have entrainment of clear air masses from the free troposphere into the ABL. The entrainment process is temporarily variable and leads locally to considerable fluctuations in the aerosol concentration. Therefore the maximum in the vertical profile of the variance of the optical backscatter intensity can be an indicator for an entrainment layer on top a CBL (Hooper and Eloranta 1986; Piironen and Eloranta 1995). The method is called variance centroid method in Menut et al. (1999). The variance method for the CBL height is also described in Lammert and Bösenberg (2006). Due to the assumptions made this method is suitable for daytime convective boundary layers only. An elucidating comparison between the gradient method and the variance method can be found in Martucci et al. (2004) although they used a Nd:YAG LIDAR at 532 nm instead of a ceilometer and thus suffered from a high lowest range gate in the order of 300 m.

B.3 RASS

The acoustic and optical methods for the determination of the mixing height, which have been described so far, are all indirect methods that try to infer the mixing height from other variables which usually adapt to the vertical structure of the ABL. The only direct and key variable for the analysis of the presence of a mixed layer is the vertical profile of virtual temperature. Temperature profiles can directly be measured with a radio-acoustic sounding system (RASS). There is also the option to derive vertical temperature profiles from Raman-LIDAR soundings (Cooney 1972) and passive radiometer measurements but especially from passive remote sensing the vertical resolution is usually not sufficient for boundary-layer research.

MLH can be determined from the lowest height where the vertical profile of potential temperature increases with height indicating stable thermal stratification of the air. The great advantage of RASS measurements is that the magnitude of stability (inversion strength) can be assessed quantitatively which is not possible from the acoustic and optical sounding devices described before.

Ideally, thermal stratification of air should be analyzed from the virtual potential temperature ($\theta_v = \theta\ (1 + 0.609\ q)$), where q is specific humidity) in order to include the effects of the vertical moisture distribution on the atmospheric stability. Unfortunately, no active remote sensing device for the determination of high-resolution moisture profiles is available. Therefore, the acoustic potential temperature ($\theta_a = \theta\ (1 + 0.513\ q)$), which actually is the temperature that is delivered by a RASS, is often used as a substitute. This is sufficient for cold and dry environments, but somewhat underestimates the virtual potential temperature in humid and warm environments. In case of larger vertical moisture gradients and small vertical temperature gradients this can lead to a switch in stability from stable to unstable or vice versa. The following two subchapters give two examples where RASS has been used for MLH determination.

B.3.1 Combined Deployment of Two Different RASS

Engelbart and Bange (2002) have analyzed the possible advantages of the deployment of two RASS instruments, a SODAR-RASS (i.e., a SODAR with an electro-magnetic extension) and a high-UHF WPR-RASS (i.e., a wind profiler with an additional sound source), to derive boundary layer parameters. With these instruments, in principle, MLH can either be determined from the temperature profiles or from the electro-magnetic backscatter intensity. The latter depends on temperature and moisture fluctuations in the atmosphere. The derivation of MLH from the temperature profile requires a good vertical resolution of the profile which is mainly available only from the SODAR-RASS. But even if the inversion layer at the top of the boundary layer is thick enough, due to the high attenuation of sound waves in the atmosphere, also the 1,290 MHz-WPR-RASS used by Engelbart and Bange (2002) can measure the temperature profile only up to about 1 km. Therefore, in the case of a deeper CBL, MLH was determined from a secondary maximum of the electro-magnetic backscatter intensity which marks the occurrence of the entrainment zone at the CBL top. Thus, with this instrument combination the whole diurnal cycle of MHL is ideally monitored by interpreting the temperature profile from the SODAR-RASS at night-time and by analyzing the electro-magnetic backscatter intensity profile from the WPR-RASS during daytime.

B.3.2 Further Algorithms Using a RASS

Hennemuth and Kirtzel (2008) have recently developed a method that uses data from a SODAR-RASS and surface heat flux data. MLH is primarily detected from the acoustic backscatter intensity received by the SODAR part of the SODAR-RASS and verified from the temperature profile obtained from the RASS part of the instrument. Surface heat flux data and statistical evaluations complement this rather complicated scheme. The surface heat flux is used to identify situations with unstable stratification. In this respect this observable takes over an analogous role as the σ_w in the EARE algorithm (Sect. B.1.4). The results have been tested against radiosonde soundings. The coincidence was good in most cases except for a very low MLH at or even below the first range gate of the SODAR and the RASS.

B.4 Other Algorithms Using More Than One Instrument

Using more than one instrument for sounding can help to overcome some of the above described deficiencies (limited vertical range, limited data availability) of the individual instruments. Possible combinations are listed in the two following sections.

B.4.1 Combined Deployment of SODAR and Wind Profiler

Beyrich and Görsdorf (1995) have reported on the simultaneous usage of a SODAR and a wind profiler for the determination of MLH. For the SODAR data the ARE method was used. From the wind profiler data MLH was likewise determined from the height of the elevated signal intensity maximum (see also Angevine et al. 1994; Grimsdell and Angevine 1998; White et al. 1999). Good agreement between both algorithms was found for evolving convective boundary layers. The vertical ranges of the two instruments (50–800 m for the SODAR and 200–3,000 m for the wind profiler) allowed following the complete diurnal cycle of MLH.

B.4.2 Combined Deployment of SODAR and Ceilometer

There is an interesting difference between the schemes for the determination of MLH from acoustic and optical backscatter intensities which should be noted carefully. While the acoustic backscatter intensity itself is taken for the detection of $H1$ and $H3$ and the first derivative of this backscatter intensity for the determination of $H2$ (see Sect. B.1.4), the first and the second derivative of the optical backscatter intensity (but not the optical backscatter intensity itself) is used to determine *H*4. This discrepancy in the processing of the two backscatter intensities is due to the different scattering processes for acoustic and optical waves: Acoustic waves are scattered at atmospheric refractivity gradients and thus at temperature gradients (Neff and Coulter 1986) while optical waves are scattered at small particles. Therefore the optical backscatter intensity is proportional to the aerosol concentration itself. The MLH on the other hand, which we desire to derive from these backscatter intensities, is in both cases found at heights where we have vertical gradients of the temperature and of the aerosol concentration. Therefore, in principle, the vertical distribution of the acoustic backscatter intensity should look very much alike to the negative of the vertical distribution of the vertical gradient of the optical backscatter intensity.

Simultaneous measurements with different remote sensing devices have mainly been made in order to evaluate one remote sensing method against the other (Devara et al. 1995). But one could also think of combining the results two or more remote sensing devices for determining the structure of the ABL. Direct detection of MLH from acoustic backscatter intensities is limited to the order of about 1 km due to the rather high attenuation of sound waves in the atmosphere. In contrast, optical remote sensing offers much larger height ranges of at least several kilometres, because the attenuation of light waves in the atmosphere is small unless there is fog, clouds or heavy precipitation.

References

Angevine W., White A.B., Avery S.K.: Boundary layer depth and entrainment zone characterization with a boundary layer profiler. Bound.-Lay. Meteorol. 68, 375–385 (1994)

Asimakopoulos D.N., Helmis C.G., Michopoulos J.: Evaluation of SODAR methods for the determination of the atmospheric boundary layer mixing height. Meteor. Atmos. Phys. 85, 85–92 (2004)

Beyrich. F., Görsdorf, U.: Composing the diurnal cycle of mixing height from simultaneous SODAR and Wind profiler measurements. Bound.-Lay. Meteorol. 76, 387-394 (1995)

Beyrich, F.: Mixing height estimation from sodar data – a critical discussion. Atmos. Environ. 31, 3941-3954 (1997)

Beyrich, F.: Mixing height estimation in the convective boundary layer using sodar data. Bound.-Lay. Meteorol. 74, 1-18 (1995)

Boers, R., Spinhirne, J.D., Hart, W.D.: Lidar Observations of the Fine-Scale Variability of Marine Stratocumulus Clouds. J. Appl. Meteorol. 27, 797–810 (1988)

Böttcher, F., S. Barth, J. Peinke: Small and large scale fluctuations in atmospheric wind speeds. Stoch. Environ. Res. Risk Assess. 21, 299–308 (2007)

Brooks, I.M.: Finding boundary layer top: application of a wavelet covariance transform to lidar backscatter profiles. J. Atmos. Oceanic Technol. 20, 1092-1105 (2003)

Carter, D.J.T.: Estimating extreme wave heights in the NE Atlantic from GEOSAT data. Health and Safety Executive – Offshore Technology Report. Her Majesty's Stationary Office OTH 93 396. 28 pp. (1993)

Cohn, S.A., Angevine, W.M.: Boundary Layer Height and Entrainment Zone Thickness Measured by Lidars and Wind-Profiling Radars. J. Appl. Meteorol. 39, 1233–1247 (2000)

Contini, D., Cava, D., Martano, P., Donateo, A., Grasso, F.M.: Comparison of indirect methods for the estimation of Boundary Layer height over flat-terrain in a coastal site. Meteorol. Z. **18**, 309-320 (2009)

Cook, N.J.: Towards better estimation of extreme winds, J. Wind Eng. Ind. Aerodyn. 9, 295-323 (1982)

Cooney, J.: Measurement of atmospheric temperature profiles by Raman backscatter. J. Appl. Meteorol. 11, 108–112 (1972)

Davis, F.K., H. Newstein: The Variation of Gust Factors with Mean Wind Speed and with Height. J. Appl. Meteor. 7, 372-378 (1968)

Davis, K.J., Gamage, N., Hagelberg, C.R., Kiemle, C., Lenschow, D.H., Sullivan, P.P.: An objective method for deriving atmospheric structure from airborne lidar observations. J. Atmos. Oceanic Technol. 17, 1455-1468 (2000)

de Haij, M., Wauben, W., Klein Baltink, H.: Determination of mixing layer height from ceilometer backscatter profiles. In: Slusser JR, Schäfer K, Comerón A (eds) Remote Sensing of Clouds and the Atmosphere XI. Proc. SPIE 6362, 63620R (2006)

Devara, P.C.S., Ernest, Ray P., Murthy, B.S., Pandithurai, G., Sharma, S., Vernekar, K.G.: Intercomparison of Nocturnal Lower-Atmospheric Structure Observed with LIDAR and SODAR Techniques at Pune, India. J. Appl. Meteorol. 34, 1375-1383 (1995)

Emeis, S., Jahn, C., Münkel, C., Münsterer, C., Schäfer, K.: Multiple atmospheric layering and mixing-layer height in the Inn valley observed by remote sensing. Meteorol. Z. 16, 415-424 (2007a)

Emeis, S., K. Baumann-Stanzer, M. Piringer, M. Kallistratova, R. Kouznetsov, V. Yushkov: Wind and turbulence in the urban boundary layer – analysis from acoustic remote sensing data and fit to analytical relations. Meteorol. Z. **16**, 393-406 (2007b)

Emeis, S., M. Türk: Wind-driven wave heights in the German Bight. Ocean Dyn. 59, 463–475. (2009)

Emeis S., Schäfer K., Münkel C.: Surface-based remote sensing of the mixing-layer height – a review. Meteorol. Z. 17, 621-630 (2008)

Emeis, S., Türk, M.: Frequency distributions of the mixing height over an urban area from SODAR data. Meteorol. Z. 13, 361-367 (2004)

Emeis, S.: Surface-Based Remote Sensing of the Atmospheric Boundary Layer. Series: Atmospheric and Oceanographic Sciences Library, Vol. 40. Springer Heidelberg etc., X+174 pp. (2011)

Engelbart, D.A.M., Bange, J.: Determination of boundary-layer parameters using wind profiler/ RASS and sodar/RASS in the frame of the LITFASS project. Theor. Appl. Climatol. 73, 53-65 (2002)

Eresmaa, N., Karppinen, A., Joffre, S.M., Räsänen, J., Talvitie, H.: Mixing height determination by ceilometer. Atmos Chem Phys 6: 1485–1493 (2006) available from: www.atmos-chem-phys.net/6/1485/2006/

Flamant, C., Pelon, J., Flamant, P.H., Durand, P.: Lidar determination of the entrainement zone thickness at the top of the unstable marin atmospheric boundary-layer. Bound.-Lay. Meteorol. 83, 247–284 (1997)

Foken, T.: Micrometeorology. Springer, 308 pp. (2008)

Gomes, L., Vickery, B.J.: On the Prediction of Extreme Wind Speeds from the Parent Distribution. J. Industr. Aerodyn. 2, 21-36 (1977)

Grimsdell, A.W., Angevine, W.M.: Convective Boundary Layer Height Measurement with Wind Profilers and Comparison to Cloud Base. J. Atmos. Oceanic Technol. 15, 1331–1338 (1998)

Gumbel, E.J.: Statistics of extremes. Columbia University Press, New York and London, 375 pp. (1958)

Hayden, K.L., Anlauf, K.G., Hoff, R.M., Strapp, J.W., Bottenheim, J.W., Wiebe, H.A., Froude, F.A., Martin, J.B., Steyn, D.G., McKendry, I.G.: The Vertical Chemical and Meteorological Structure of the Boundary Layer in the Lower Fraser Valley during Pacific'93. Atmos. Environ. 31, 2089–2105 (1997)

Hennemuth, B., Kirtzel, H.-J.: Towards operational determination of boundary layer height using sodar/RASS soundings and surface heat flux data. Meteorol. Z. 17, 283-296 (2008)

Hooper, W.P., Eloranta, E.: Lidar measurements of wind in the planetary boundary layer: the method, accuracy and results from joint measurements with radiosonde and kytoon. J. Clim. Appl. Meteorol. 25, 990-1001 (1986)

Jensen, N.O., L. Kristensen: Gust statistics for the Great Belt Region. Risoe-M-2828, 21 pp. (1989)

Justus, C.G., W.R. Hargraves, A. Mikhail, D. Graber: Methods for Estimating Wind Speed Frequency Distributions. J. Appl. Meteor. 17, 350-353 (1978)

Justus, C.G., W.R. Hargraves, A. Yalcin: Nationwide Assessment of Potential Output fromWind-Powered Generators. J. Appl. Meteor. 15, 673-678 (1976)

Kaimal, J.C., S.F. Clifford, R.J. Lataitis (1989) Effect of finite sampling on atmospheric spectra. Bound-Lay Meteorol 47, 337-347

Lammert, A., Bösenberg, J.: Determination of the Convective Boundary-Layer Height with Laser Remote Sensing. Bound.-Lay. Meteorol. 119, 159-170 (2006)

Martucci, G., Srivastava, M.K., Mitev, V., Matthey, R., Frioud, M., Richner, H.: Comparison of lidar methods to determine the Aerosol Mixed Layer top. In: Schäfer K, Comeron A, Carleer M, Picard RH (eds.): Remote Sensing of Clouds and the Atmosphere VIII. Proc of SPIE 5235, 447-456 (2004)

Melfi, S.H., Spinhirne, J.D., Chou, S.H., Palm, S.P.: Lidar Observation of the Vertically Organized Convection in the Planetary Boundary Layer Over the Ocean. J. Clim. Appl. Meteorol. 24, 806–821 (1985)

Menut, L., Flamant, C., Pelon, J., Flamant, P.H.: Urban Boundary-Layer Height Determination from Lidar Measurements Over the Paris Area. Appl. Opt. 38, 945-954 (1999)

Mitsuta, Y., O. Tsukamoto: Studies on Spatial Structure of Wind Gust. J. Appl. Meteor. 28, 1155-1161 (1989)

Morales, A., M. Wächter, J. Peinke: Advanced characterization of wind turbulence by higher order statistics. Proc. EWEC 2010 (2010)

Münkel, C., Räsänen, J.: New optical concept for commercial lidar ceilometers scanning the boundary layer. Proc. SPIE 5571, 364–374 (2004)

Münkel, C.: Mixing height determination with lidar ceilometers – results from Helsinki Testbed. Meteorol. Z. 16, 451-459 (2007)

Neff, W.D., Coulter, R.L.: Acoustic remote sensing. In: Lenschow DH (Ed.) Probing the Atmospheric Boundary Layer. Amer Meteor Soc, Boston, MA, 201–239 (1986)

Palutikof, J.P., B.B. Brabson, D.H. Lister, S.T. Adcock: A review of methods to calculate extreme wind speeds. Meteorological Applications, 6, 119-132 (1999)

Panchang, V., Zhao, L., Demirbilek. Z.: Estimation of extreme wave heights using GEOSAT measurements. Ocean Eng. 26, 205-225 (1999)

Piironen, A.K., Eloranta, E.W.: Convective boundary layer depths and cloud geometrical properties obtained from volume imaging lidar data. J. Geophys. Res. 100, 25569-25576 (1995)

Schäfer, K., Emeis, S., Junkermann, W., Münkel, C.: Evaluation of mixing layer height monitoring by ceilometer with SODAR and microlight aircraft measurements. In: Schäfer K, Comeron AT, Slusser JR, Picard RH, Carleer MR, Sifakis N (eds) Remote Sensing of Clouds and the Atmosphere X. Proc. SPIE 5979, 59791I-1 – 59791I-11 (2005)

Schäfer, K., Emeis, S.M., Rauch, A., Münkel, C., Vogt, S.: Determination of mixing-layer heights from ceilometer data. In: Schäfer K, Comeron AT, Carleer MR, Picard RH, Sifakis N (eds.): Remote Sensing of Clouds and the Atmosphere IX. Proc. SPIE 5571, 248-259 (2004)

Schroers, H., H. Lösslein, K. Zilich: Untersuchung der Windstruktur bei Starkwind und Sturm. Meteorol. Rdsch. 42, 202-212 (1990)

Seibert, P., Beyrich, F., Gryning, S.-E., Joffre, S., Rasmussen, A., Tercier, P.: Review and intercomparison of operational methods for the determination of the mixing height. Atmos. Environ. 34, 1001-1027 (2000)

Senff, C., Bösenberg, J., Peters, G., Schaberl, T.: Remote Sesing of Turbulent Ozone Fluxes and the Ozone Budget in the Convective Boundary Layer with DIAL and Radar-RASS: A Case Study. Contrib. Atmos. Phys. **69**, 161–176 (1996)

Sicard, M., Pérez, C., Rocadenbosch, F., Baldasano, J.M., García-Vizcaino, D.: Mixed-Layer Depth Determination in the Barcelona Coastal Area From Regular Lidar Measurements: Methods, Results and Limitations. Bound.-Lay. Meteorol. 119, 135-157 (2006)

Steyn, D.G., Baldi, M., Hoff, R.M.: The detection of mixed layer depth and entrainment zone thickness from lidar backscatter profiles. J. Atmos. Ocean Technol. 16, 953–959 (1999)

Türk, M.: Ermittlung designrelevanter Belastungsparameter für Offshore-Windkraftanlagen. PhD thesis University of Cologne (2008) (Available from: http://kups.ub.uni-koeln.de/2799/)

Van der Hoven, I.: Power Spectrum of Horizontal Wind Speed in the Frequency Range from 0.0007 to 900 Cycles per Hour. J. Meteorol. 14, 160-164 (1957)

White, A.B., Senff, C.J., Banta, R.M.: A Comparison of Mixing Depths Observed by Ground-Based Wind Profilers and an Airborne Lidar. J. Atmos. Oceanic Technol. 16, 584–590 (1999)

Wieringa, J (1973) Gust factors over open water and built-up country. Bound-Lay Meteorol 3: 424-441

Wieringa, J.: Shapes of annual frequency distributions of wind speed observed on high meteorological masts. Bound.-Lay.Meteorol. 47, 85–110 (1989)

Wulfmeyer, V., Janjić, T.: 24-h observations of the marine boundary layer using ship-borne NOAA High-Resolution Doppler Lidar. J. Appl. Meteorol. 44, 1723-1744 (2005)

Wulfmeyer, V.: Investigation of turbulent processes in the lower troposphere with water-vapor DIAL and Radar-RASS. J. Atmos. Sci. 56, 1055-1076 (1999)

Index

S. Emeis, *Wind Energy Meteorology*, Green Energy and Technology,
DOI: 10.1007/978-3-642-30523-8,

O

P

R

S

T

U

V

W

Printed by Publishers' Graphics LLC
CAMZ130812.15.18.181